T0296254

CAMBRIDGE LIBRARY COLLECTION

Books of enduring scholarly value

Life Sciences

Until the nineteenth century, the various subjects now known as the life sciences were regarded either as arcane studies which had little impact on ordinary daily life, or as a genteel hobby for the leisured classes. The increasing academic rigour and systematisation brought to the study of botany, zoology and other disciplines, and their adoption in university curricula, are reflected in the books reissued in this series.

Journal of a Tour in Iceland, in the Summer of 1809

Sir William Jackson Hooker (1785–1865) was an eminent British botanist, best known for expanding and developing the Royal Botanic Gardens at Kew into a leading centre of botanic research and conservation. At the age of nineteen he undertook an expedition to Iceland, his first outside Britain. Unfortunately, all his specimens and notes were destroyed in a fire on the return voyage (described in Volume 1), but he was able, with the help of the notes made by Sir Joseph Banks on an earlier expedition, to write this account. His work was first published privately in 1811, but a second edition was published in 1813 and is reproduced here. Volume 1 gives a brief history of Iceland, before Hooker begins his detailed observations of the people and topography, and the flora and fauna he found. His accounts of the lives of the people of the island are of particular interest.

Cambridge University Press has long been a pioneer in the reissuing of out-of-print titles from its own backlist, producing digital reprints of books that are still sought after by scholars and students but could not be reprinted economically using traditional technology. The Cambridge Library Collection extends this activity to a wider range of books which are still of importance to researchers and professionals, either for the source material they contain, or as landmarks in the history of their academic discipline.

Drawing from the world-renowned collections in the Cambridge University Library, and guided by the advice of experts in each subject area, Cambridge University Press is using state-of-the-art scanning machines in its own Printing House to capture the content of each book selected for inclusion. The files are processed to give a consistently clear, crisp image, and the books finished to the high quality standard for which the Press is recognised around the world. The latest print-on-demand technology ensures that the books will remain available indefinitely, and that orders for single or multiple copies can quickly be supplied.

The Cambridge Library Collection will bring back to life books of enduring scholarly value (including out-of-copyright works originally issued by other publishers) across a wide range of disciplines in the humanities and social sciences and in science and technology.

Journal of a Tour in Iceland, in the Summer of 1809

VOLUME 1

WILLIAM JACKSON HOOKER

CAMBRIDGE UNIVERSITY PRESS

Cambridge, New York, Melbourne, Madrid, Cape Town,
Singapore, São Paolo, Delhi, Tokyo, Mexico City

Published in the United States of America by Cambridge University Press, New York

www.cambridge.org
Information on this title: www.cambridge.org/9781108030489

This edition first published 1813
This digitally printed version 2011

ISBN 978-1-108-03048-9 Paperback

AN ICELANDIC LADY IN HER BRIDAL DRESS.

JOURNAL

OF

A TOUR IN ICELAND,

IN

THE SUMMER

OF

1809.

BY

WILLIAM JACKSON HOOKER, F. L. S.,

AND

FELLOW OF THE WERNERIAN SOCIETY OF
EDINBURGH.

SECOND EDITION, WITH ADDITIONS.

VOL. I.

LONDON:

PRINTED FOR LONGMAN, HURST, REES, ORME, AND BROWN,
PATERNOSTER-ROW,
AND JOHN MURRAY, ALBEMARLE-STREET,

By J. Keymer, Yarmouth.

1813.

TO

THE RIGHT HONORABLE

SIR JOSEPH BANKS, BART., K. B.,

PRESIDENT OF THE ROYAL SOCIETY,

&c., &c., &c.

My Dear Sir,

I feel a peculiar propriety in dedicating this little work to you, and, unworthy as it is in itself of the honor of being sent into the world under the sanction of a name like yours, I trust that you will not refuse to accept it as a proof of the esteem and respect of the author. I have two particular reasons for being anxious it should thus appear: the one, because it is right that the earliest efforts of my pen should be inscribed to him, who, by proposing and facilitating my *Tour to Iceland*, first gave that pen the opportunity of being employed; the other, because it is chiefly in obedience to your advice that I now lay before the public what was originally written for the perusal only of my personal friends. These friends have, indeed, done me the kindness to receive this book in a manner the most gratifying to me; but the partiality of friends is proverbial, and in the public I must expect to meet with less favorable judges: the apprehension, therefore, which I cannot but feel of their criticism at my first appearance before them, makes me desirous to shield myself under the authority of a man, to whose judgment they are accustomed to pay the same deference that I do. As a farther reason for the change of my intention, I must be allowed to alledge the circumstance, that I found my own withholding this book would not prevent its actual publication; different parts of it having already appeared in periodical works, which have announced their intention of continuing similar extracts; and I consequently considered it more respectful to the public, if not due to myself, that, such as it is, they should have the opportunity of perusing it entire, instead of having it forced upon their attention in garbled extracts.

I have the honor to be,

MY DEAR SIR,

Your obliged friend,

and very humble Servant,

W. J. HOOKER.

LONDON,
10*th* AUGUST, 1811.

PREFACE.

THE interesting Letters on Iceland, published by the Archbishop Von Troil, had inspired me at an early age with an ardent desire to visit that most singular country, to see its volcanic mountains and its boiling springs, unequalled by any in the known world. The small degree of intercourse, however, that existed between England and so obscure a part of the globe, and, what appeared a still greater impediment, the unfriendly disposition exhibited by the Danish government towards our nation, scarcely allowed even an hope to be entertained that an opportunity of gratifying such a wish would present itself, till, in the spring of the year 1809, Sir Joseph Banks most unexpectedly proposed to me, as a compensation for my not having it in my power, during that season, to put in execution a projected voyage to a tropical climate, that I should take my passage in a merchant-ship, then expected to leave England in the space of three days, and spend my summer in Iceland. To this I most readily and thankfully acceded; and, having made such preparations as the shortness of the notice would allow, I repaired to Gravesend and was on board the Margaret and Anne at the time appointed.

The unfortunate accident, which has deprived me of nearly the whole of the fruits of this excursion, and has obliged me to rely, in no small degree, upon my memory, needs not here to be detailed, it will find its place in the narrative of the voyage; suffice it now to observe, that the only things rescued from the flames were, a portion of my journal, containing little more than the occurrences of the first four weeks of my stay on the island, and an Icelandic lady's wedding-dress, which was saved by the extraordinary exertions of the steward of the ship. Of the rest of my manuscripts and collections, including plants, books, drawings, minerals, and other subjects of natural history, nothing could be preserved.

With the slender materials that remained to me, I should not have ventured upon committing the following Recollections to paper, even as they were originally intended, merely for the perusal of some of my friends, but at the solicitation of the most valued of those friends. It is to Mr. Dawson Turner, of Yarmouth, that these sheets owe their existence.

To Sir Joseph Banks, besides being honored with his counsel and assistance preparatory to the undertaking of the voyage, I am indebted for the truly hospitable entertainment that I experienced from the inhabitants of Iceland, who felt, I am sure, a real pleasure and satisfaction in having it in their power to offer their services and to pay every possible attention to a stranger, visiting their country with an introduction from their great and generous benefactor. Not, however, satisfied with this, on my return to England, no sooner did Sir Joseph learn that I was preparing my *Recollections of*

Iceland for the press, than he most liberally offered me the use of his own manuscript journal, and various other papers and documents relative to the island, together with the magnificent drawings of the scenery, dresses of the inhabitants, &c., which were made by the artists who attended him on his voyage thither, in 1772. From the former of these valuable collections I have extracted such parts as were not noticed by Von Troil; and, from reduced copies of a few of the latter, have been made the engravings that accompany these volumes. These are, indeed, upon too small a scale to give an adequate idea of the originals, which would do honor to a large and copious history of Iceland; but *parvum parva decent*, and they are well suited both to the size and pretensions of the book they are designed to illustrate.

The reception which I met with from the merchants and owners of the vessel in which I sailed, Messrs. Phelps, Troward, and Bracebridge, and the assistance which I derived from them, demand my most sincere thanks; the readiness with which the former of these gentlemen, in particular, whose society I enjoyed during the voyage, entered into all my views, and the willingness with which he supplied me with every thing that could afford me accommodation, or might further the object of my pursuits, have left a lasting impression of gratitude upon my mind.

Neither can I suffer to pass in silence the civility of Sir George Mackenzie, in collecting plants for me in his late excursion to Iceland; nor the attention shown me by Doctor Wright, of Edinburgh. Though a stranger to the latter gentleman, till my arrival in Scotland on my return from Iceland, he nevertheless participated feelingly in my misfortunes, and begged me to make any use I pleased of the various subjects of natural history in his possession, which had been collected in Iceland by his nephew, the late Mr. Wright, an amiable young man, who accompanied Sir John Stanley on his voyage to that country.

No apology, I trust, will be considered necessary for prefacing my journal with a slight and very cursory sketch of Icelandic history, or with the details that follow, explanatory of the various offices, as well civil as ecclesiastical. An introduction, comprising these, and hints on a few other most remarkable objects in the island, appeared to me to be necessary, not only for the proper understanding of much of my narrative, but to prevent these volumes from being to such a degree incomplete as would have rendered them almost useless.

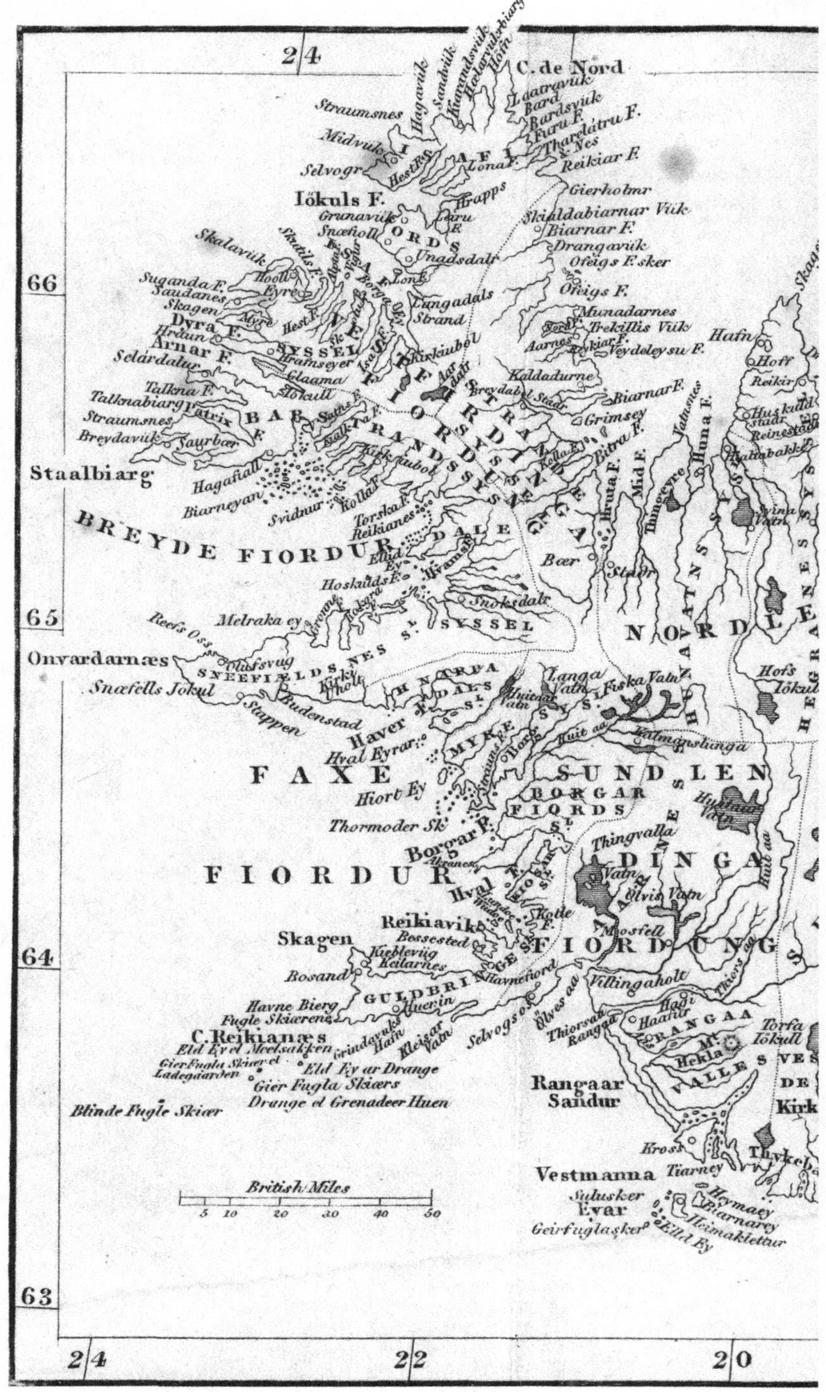
24
C. de Nord
Straumsnes
Midvik
Selvogr
Iökuls F.
Skalavik
Suganda F.
Saudanes
Skagen
Dyra F.
Arnar F.
Selárdalur
Talkna F.
Talknabiarg
Patrix F.
Straumsnes
Breydavik
Saurbær
Staalbiarg
Hagafiall
Biarneyan
BREYDE FIORDUR
Melraka ey
Onvardarnæs
Snæfells Jökul
Budenstad
Stappen
Haver
Hval Eyrar
FAXE
Hiort Ey
Thormoder Sk
FIORDUR
Borgar F.
Hval F.
Reikiavik
Bessested
Skagen
Kieblevig
Keilarnes
Rosand
GULDBRIN
Havne Bierg
Fugle Skiærene
C. Reikianæs
Eld Ey el Meelsakken
Eld Ey ar Drange
Gier Fugla Skiærs
Drange el Grenadeer Huen
Blinde Fugle Skiær
Gierholmr
Skialdabiarnar Vik
Biarnar F.
Drangavik
Ofeigs F. sker
Ofeigs F.
Munadarnes
Trekillis Vik
Veydeleysu F.
Kaldadurne
Biarnar F.
Grimsey
Bitra F.
Hruta F.
Mid F.
Huna F.
Hafn
Hoff
Reikir
Bær
Stadr
NORDLEN
Snoksdalr
SYSSEL
Fiska Vatn
Langa Vatn
Huit aa
SUNDLEN
BORGAR FIORDS
Thingvalla Vatn
FIORDUNG
Mosfell
Villingaholt
Olves aa
Selvogsos
Thiorsaa
Rangaa
RANGAA
Mt. Hekla
Torfa Iökull
Rangaar Sandur
Kross
Tiarney
Vestmanna Eyar
Hrymaey
Biarnarey
Heimaklettur
Eld Ey
Geirfuglasker
Sulusker
Hofs Iökul
British Miles
5 10 20 30 40 50
66
65
64
63
24
22
20

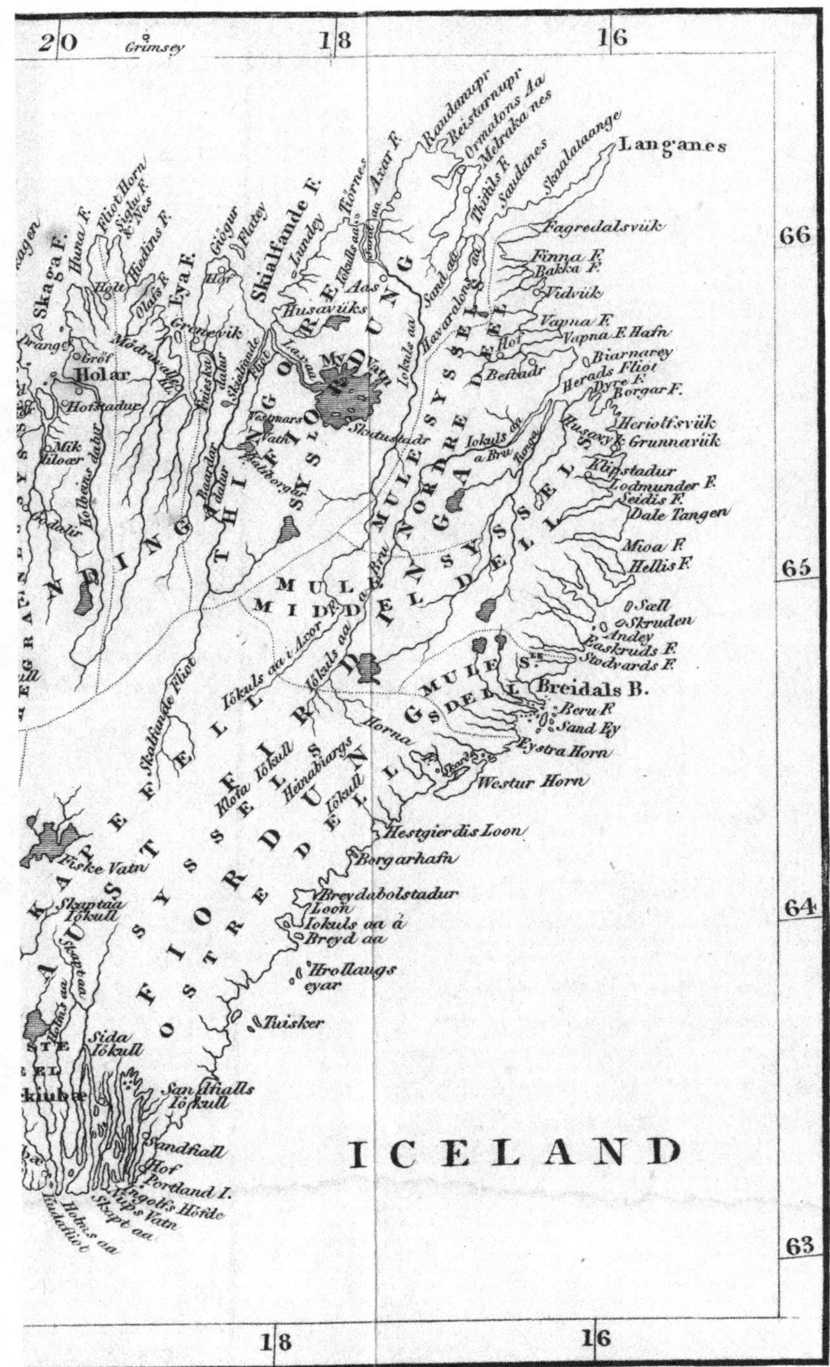

ICELAND
Langanes
Fagredalsvük
Finna F.
Bakka F.
Vidvük
Vapna F.
Vapna F. Hafn
Biarnarey
Herads Fliot
Borgar F.
Heriolfsvük
Grunnavük
Klipstadur
Lodmunder F.
Seidis F.
Dale Tangen
Mioa F.
Hellis F.
Sæll
Skruden
Andey
Faskruds F.
Stodvards F.
Breidals B.
Beru F.
Sand Ey
Eystra Horn
Westur Horn
Hestgierdis Loon
Borgarhafn
Breydabolstadur
Loon
Iokuls aa á
Breyd aa
Hrollaugs eyar
Tuisker
Sandfialls Iökull
Sandfiall
Hof
Portland I.
Ingolfs Höfde
Skaptaa Iökull
Fiske Vatn
Sida Iökull
Holar
Hofstadur
Grenevik
Husavüks
Skialfande F.
Skialfande Fliot
Grimsey
Skaga F.
Eya F.
Hof
Flatey
Skutustadr
Besstadr
Horna
Heinabiargs
Klofa Iökull
Iökuls aa i Axor
Raudanupr
Reistarnupr
Ormatons Aa
Melraka nes
Thistils F.
Saudanes
Skaalataange
Axar F.
Tiörnes
Lundey
Aas
20
18
16
66
65
64
63

INTRODUCTION.

ICELAND, one of the most considerable of the European isles, is situated in the northern part of the Atlantic Ocean; and, according to the most authentic observations that have been made, between the sixty-third and sixty-seventh degrees of north latitude, and the sixteenth and twenty-fifth degrees of longitude, west of Greenwich*.

* The able French navigator, M. de Verdun, de la Crenne, whilst speaking of the maps of Iceland in his *Voyage en diverses Parties de l'Europe de l'Afrique et de l'Amerique,* takes the opportunity of remarking " qu' elles diffèrent tant entr' elles, et d'ailleurs elles s'accordent si peu avec le résultat de nos observations, par rapport à la partie, que nous avons parcourue, qu' il ne nous est pas possible de présumer qu' elles representent plus fidèlement les parties que nous n'avons pu reconnôitre." — Arngrim Jonas, likewise,

Whether or not this island was the Thule of any of the ancient writers, who have spoken of a country by that name, is a question which has been often discussed, and by

in his *Brevis Commentarius de Islandiâ*, after having, with great pains, collected many of the various opinions that existed, in his time, respecting the situation of Iceland, observes, to use the words of his translator, "There be others, also, who either in their maps or writings have noted the situation of Iceland; notwithstanding it is to no purpose to set down any more of their opinions, because the more you have the more contrary shall you find them." M. de Verdun took a very accurate observation in the middle of the Westmann's Isles, which lie very near to, and correspond with, the southern point of the main land, and the position given was 63° 20′ 30″ of north latitude, and 22° 47′ 50″ longitude, west of Paris. From another observation made by the same officer at Cap Nord, the most northern point of the island, its situation was ascertained to be in 66° 44′ north longitude, and 25° 4′ latitude, west of Paris.—With regard to the maps which accompany the present publication, Mr. Arrowsmith has, with great care and assiduity, collected information from the best authorities, in order to render them the most correct of any that have yet appeared; and he has not only made ample use of the volumes of M. Verdun, but also of manuscript maps and charts which have been constructed by Danish officers, who have been employed in Iceland at the expence of his Danish Majesty.

none perhaps more ably than by Arngrim Jonas, in his *Tractatus de Islandiâ;* but it nevertheless remains undecided. Still more to be doubted are the accounts that have been handed down to us, principally by Geoffrey, of Monmouth, of the subjugation* of the island by King Arthur, and of the subsequent arrival in England of a King† of Iceland to do homage to that prince. Were these particulars substantiated, the relation of the discovery and colo-

* "Anno Christi 517, Arthurus, secundo regni sui anno, subjugatis totius Hiberniæ partibus, classem suam direxit in Islandiam, eamque, debellato populo, subjugavit."— *Galfrid. Monumet. Hist. Briton, lib.* 9. c. 10.

† "Missis deinde in diversa regna legatis, invitantur tàm ex Galliis, quàm ex collateralibus autem insulis oceani, qui ad curiam venire deberent; ex collateralibus autem, insulis, Guillaumurius rex Hiberniæ, *Maluasius rex Islandiæ,* Doldavius rex Gotlandiæ, Gunnasius rex Orcadum, Lot rex Norvegiæ, Aschilius rex Danorum."— *Galfrid. Monum. lib.* 9. c. 12.— It is further mentioned, in the nineteenth chapter of the ninth book of the same author, that Prince Arthur had six score thousand soldiers sent him from these six countries!

nization of Iceland, as given by the most respectable historians of the country, must be looked upon as a fable.

Following, therefore, the native writers of Iceland, its earliest discoverer upon record was a famous pirate of the name of Naddoc, a Norwegian by birth, who, in the year 861, was driven thither by a tempest, while on a voyage from his native country to the Ferroe Islands; and, on account of the vast quantity of snow, with which he observed the mountains to be covered, named it Snoeland. Not alarmed, however, by this chilling prospect, such was the account of the country he gave on his return home, that others were induced to go in search of it. The first of these, Gardar Suaversen, a native of Sweden, set sail in the year 864, and, after approaching the eastern coast, proceeded round the island to a harbor in the north, where he came to an anchor, and passed the winter at a place which has since borne the appellation of Skialfiord: in order to immortalize himself for this bold exploit, he altered the name of the island to Gardarsholme. The next adventurer was Floco; who, as the compass was not yet discovered, to remedy this deficiency, took in his vessel some ravens, the sacred bird of the north; one of which, at the time when he supposed he was drawing towards the termination of his voyage, he suffered to escape, hoping, by its course, to be more surely directed towards the country of which he was in search; the bird, however, turned his flight towards Haitland, the port whence they had set out, and satisfied Floco that he was still at a less distance from Norway than from Gardarsholme. Pursuing his voyage, therefore, for some time longer, he at length liberated another raven, who, finding "no rest for the sole of his feet," returned, and took refuge in the vessel. In a few days a third raven was suffered to leave the ship, and this, more fortunate, pursued its course towards the long-expected shore. Floco, in like manner as his predecessors had done, first touched at the eastern coast, whence, steering his course round the southern part of the island, he entered the great gulf (now called Faxa-fiord) between the two promontories that have since been distinguished by the names of Snoefel-nes and Reikanes; but, afterwards, proceeding northward, he harbored for the winter at Watnsfiordur, in the gulf of Breidafiord. So great was the quantity of ice which, in the spring of the following year, entered the harbor, that Floco was tempted, in consequence of it, once more to change the name of the island, and give it that which it has ever since retained. He passed another winter in the southern part of the country previously to his return to Norway, where, on account of the use he had made of the ravens, he obtained the appellation of Rafnafloke.

Induced by the relation given by Floco of the condition of the new country, Ingulf, a Norwegian, of noble birth and great opulence, having fallen under the displeasure of the tyrant, Harald Hafalgar, conceived, together with his friend, Hiorleif, the project of establishing themselves in Iceland: in pursuance of this plan, the former sailed, in 870, for the purpose of exploring its shores; but no settlement was made till the year 874, when they both emigrated, accompanied by their respective families and numerous followers. In compliance with a custom among the Norwegians, that was sanctioned by the religion of those days, Ingulf, on his approach to the coast, cast the door-posts of the house which he had left into the sea, that wheresoever they were thrown on shore he might establish his infant colony; but, being himself driven in a different direction from them, he was reduced to the necessity of landing on a promontory, which to this day bears the name of Ingulfshöfde, in the south-eastern part of the island; and it was not till after a period of three years that the posts were found on the shore of the bay where Reikevig now stands, to which spot Ingulf, with his family, immediately repaired, and built their habitation. Hiorleif, regardless of heathenish superstitions, fixed his abode at a place called Hiorleifshöfde, and employed himself and his attendants in the cultivation of the soil. A termination was soon put to his improvement and his life by some Irish servants, whom he had brought with him from Norway, and who afterwards fled to the scarcely accessible rocks of the Westmann's Isles, where Ingulf* pursued and slew them.

Iceland is said to have been so entirely overgrown with thick forests of birch, that whenever the settlers had occasion to make excursions into the country, they were forced

* The spot where Ingulf's remains were interred is pointed out to this day, and is known by the name of Ingulfshaugur: "Ce tombeau, qui consiste en une grande butte, peut être vu distinctement du canton; il a deux cents toises de circonférence, et paraît comme un tertre naturel formé de gravier, de pierres, et en partie de la roche même. Il n'est point invraisemblable que cet Ingulf soit enterré ici, la raison qu' on en donne, toute singulière qu' elle est, le confirme; Ingulf a ordonné, dit on, qu on le fit enterrer au sommet de cette montagne, afin de pouvoir dans l'autre vie, promener librement ses regards sur une vaste étendue du pays qu' il avait conquis; ce qui s'accorde fort bien avec les idées superstitieuses des payens du mord. L'évêque Brynjulf Svendsen, qui aimait beaucoup les monumens antiques, se transporta sur cette montagne, accompagné d'un des meilleurs poètes de l'islande, qui, sur la demande de l'évêque, composa sur le lieu un chant en honneur d'Ingulf; l'évêque et sa société y élevèrent en même temps, de leurs propres mains, des pierres qu' ils y trouvèrent, une pyramide, à la mémoire d'Ingulf.

to open passages with the axe. The coasts did not appear to have been at all times without inhabitants, though there is great reason to suppose that they were only casual visitors; and, from various little utensils which were found belonging to Christian worship, that they were of that religion. Are Frode observes, in the Landnama Saga, that these people were called, by the Norwegians, Papar; in whom seem to have originated the names of Papa-sound, in Norway, Papey, in Iceland, and Papay Stronsay and Papay Westray, in the Orkney Isles. The eastern coasts of the island were the most frequented by the Papar*, who are by many supposed to have been Irish priests, who labored to convey the blessings of Christianity among many of the northern nations.

So great was the number of Norwegians who now followed Ingulf, in order to escape

* The word *Papa*, or Pope, has not always been applied to the head of the Romish church, but was in ancient times given to all bishops; and it is only since Gregory VII. that it has been appropriated to the bishop of Rome. — See *Jortin's Ecclesiastical History. v. 5. p. 64.*

the yoke of an oppressor at home, that, in the course of sixty years, the whole of the coasts, and most of the habitable parts, are said to have been occupied by the new settlers.

As the population increased, the necessity of having a regular form of government became apparent; and, accordingly, about the year 928, a constitution was established, which continued in full vigor for more than three hundred and thirty years. This early state of the republic was evidently an aristocracy. The island was divided into four quarters, to which were given the names of the cardinal points of the compass; these also into three (with the exception of the northern quarter, where, on account of its extent, there were four) lesser divisions, or prefectures, which were called Pyng; and these again were subdivided into ten or more districts, called Hreppars. In each of them a number of inhabitants, not less than twenty, possessed of a certain portion of landed property*, constituted an assembly. Out of

* Those who were to be admitted into this assembly were chosen at a meeting of all the members; it was

this body, five were chosen, who were the most celebrated, in the first instance, for their wisdom and integrity, and, in the next, for their wealth and possessions (lest they should be exposed to contempt or corruption) and appointed judges, or Hreppsstiorars, who were privileged to convoke the assemblies, to pass sentence, and to punish crimes in their respective courts. It was their office, moreover, to attend to the wants of the poor, and to prevent, as much as lay in their power, the lower class of people* from

particularly their interest to exclude all such as were likely to be reduced in circumstances; for, in that case, the person and his family were maintained at the expence of the assembly. It was, moreover, enacted by law, that, when any of the members of the Hreppar were suffering under the loss of houses or buildings, by fire or any other accident, or of cattle by disease, an estimate of the damage should be made within fourteen days and the full amount be paid to the sufferer, either out of public funds, established for the purpose, or by a collection made at the house of every individual, each member paying according to his substance.

* Every pauper was to be admitted into the family of his nearest relative, if he had any, otherwise he was to reside in his own Hreppar, and be supported entirely at the public expence.

becoming burthensome to the nation. In this infant state of the community it was looked upon as most disgraceful to become a beggar through idleness. Arngrim Jonas remarks, that it was an apostolic precept, that he who would not labor should not be suffered to beg, and that consequently severe punishments* were inflicted on persons so offending, and heavy fines imposed upon those who were detected of harboring and encouraging this class of people.

The Hreppsstiorars, as has been just stated, had authority to convoke assemblies of the people; and these assemblies may be considered of two kinds, the fixed or ordinary, and the extraordinary, or such as were called on special occcasions. It appears that, at both of them, each member carried with him some emblem or sign, which, since the introduction of christianity, was a small wooden cross, and the omission of it was considered

* Of such a description, is the following: "Item altera lex de ejusmodi mendicis impunè castrandis, etiamsi cum eorundem nece conjunctum foret, ne videlicet ostiatim vivendo liberos gignerent sin.iles parentibus, qui postmodum oneri essent Reipublicæ. *Islandi Tractatus. p.* 437.

as a crime which merited punishment. Here were held consultations upon public affairs, and here accused persons were brought for trial and to receive punishment. If the complaint was of a private nature, the accuser himself summoned the defendant; or, if he was unable or unwilling to do it, one of the Hreppsstiorars undertook the office for him. It was his duty, likewise, to bring forward all public acts of injustice, but, should he be negligent in the exercise of his functions, he was subject to be called to an account by the other members of the Hreppar, and mulcted in a pecuniary fine.

Each quarter of the island, as has been already stated, contained three (except the northern, which had four) lesser divisions or præfectures. These were under the superintendence of magistrates of much greater rank and dignity than the Hreppsstiorars, and may be looked upon as the Præfects * of

* These were the nobles of the land: "Eos Optimates hoc loco appellamus, et statum Reipublicæ, horum inspectione gubernatum, Aristocratiam." *Arngrim Jonas Islandi Tract.*

the district in which they resided. Their influence extended to matters relating to ecclesiastical as well as to civil affairs. They were the priests in the sacred places, and judges in business relating to the law; although it must be observed, that the interpretation of the law belonged, in a more particular manner, to a person of still higher authority, hereafter to be noticed. The Præfects were, in the Icelandic, denominated *Godar* or *Hoffgodar* (from *Hoff* a temple), and their office was called *Godord.* In order to give the greater dignity to their meetings, they were convened in the sacred places, and, in like manner as the Reppstiorars, may be looked upon as of two kinds, the ordinary or annual, and the extraordinary, or those that were appointed only in cases of emergency. Each of them was distinguished by some sign or emblem. As the head of the church, within his own præfecture, was part of the office of this magistrate to appoint the sacrifices and ceremonies that were to be performed in the temples; to collect the tribute-money for the expences attending religious worship and keeping the sacred buildings in repair; as well as to impose fines* on those who were found guilty of profaning the temples or speaking irreverently of the gods.

When any affair occurred of great importance, or such as concerned the whole province, the three Præfects of such province assembled, and formed the Fiordnugathyng, or States of the Quarter. These were as often convoked as any danger threatening the whole province seemed to require, or the quarrels among the different communities of the præfectures rendered necessary.

Superior, however, to all the magistrates that have now beeen described, was the Logmann, or Logsogmann, who was elected, by the choice of the people, sovereign Judge of the whole Island. He was, as his title implies, the expounder of the law. He

* In Iceland and Norway all crimes were rated at a certain number of marks. The mark was divided into eight parts, each of which was equivalent to six ells of wadmal; consequently one mark (which consisted of somewhat more than an ounce of fine silver) was equal in value to forty-eight ells of this cloth.

enacted new laws, annulled or changed the old ones, and was charged with seeing them put in execution; and when written laws came into use, the Logmann had them in his custody. This magistrate chiefly officiated in the great assembly or Althing, which he convoked annually, and which was attended by every member of the state and by every citizen of the island. Here the more weighty and important causes were brought forward; and the provincial judges were induced to conduct themselves, in their respective jurisdictions, with the greater caution, lest their acts should be represented to this assembly and they thereby be subject to be condemned and punished; for to this court lay an appeal from the sentences pronounced in all the inferior courts. This great assembly of the states, which was always begun and ended with sacrifices, lasted fourteen days, beginning in the month of May; and was held, for some time, at Armanfel in the southern part of the island; but afterwards at Thingevalle. The Icelandic historians have with great care preserved the names of those persons in the island, who have been elevated to the rank of Logmenn, and, by the list of them published by Arngrim Jonas, we learn that there were in all thirty-eight. Among those most deserving of notice, Rafnerus, the son of Cetellus Hange, may be mentioned as the first who was constituted Logmann, in Iceland, in the year 930; Thorgeirus Lioswetninga Bode, during whose reign paganism was abolished, in 1000; Bergthorus, who established the canon law; Snorro Sturleson, the famous historian and poet, who was chosen in 1215; and Cetillus or Catullus, the last of the Logmenn, who maintained his authority from 1259 to 1262, at which time, having long withstood the threats and solicitations of Haco, king of Norway, it was agreed by the Icelanders, in a national assembly, that they should do homage to that prince; and they accordingly became the subjects of Norway, after having maintained their independence for upwards of three hundred and thirty years. Although a Norwegian governor was appointed to reside in the island, it does not appear that the internal constitution, or the laws, underwent any material change. The people continued faithful in their allegiance to their new masters, and became, with them, subservient to the crown of Denmark, in the year 1387.

The code of laws, called the Jonsbok, was received in Iceland in 1280; but this seems to have been principally founded on the more ancient laws of the island. It underwent much alteration when the Danes had possession of the country, till, at length, most matters were decided by the law of Denmark, and continue to be so to this day, with some few exceptions and alterations, adapted to local circumstances.

The Danes have entrusted the government* of the island to a person who is styled Stiftsamptman, that is, the supreme governor of a province or stiftsampt. The stiftsampt of Iceland is divided into four ampts, each of which was formerly under

* For the account of the present state of the civil as well as ecclesiastical establishments contained in this Introduction, as well as for that of the state of commerce of the island, I am greatly indebted to Mr. Jorgensen, who, from the late situation he held in Iceland, has been no less able than willing to furnish me with much useful information.

the care of an Amptman, who is a sort of deputy governor and the second magistrate in the island; but at this time there are but two of these; the southern ampt having been put under the immediate cognizance of the Stiftsamptman, and the eastern one united with the northern.

The ampts are again divided into about twenty syssels, and these into repps. To each syssel is prefixed a Sysselman, whose office it is to collect the royal revenues, either in kind or money, according to the regulation of each particular district. They all receive their salaries out of the taxes, excepting only one or two, who are paid an annual sum by the Landfogued.

A repp is superintended by a person called Reppstiorar, who is subordinate to the Sysselman, as the latter is to the Amptman, and whose duty, besides that of seeing to the peace and good order of the community, is in a particular manner directed to the care and maintenance of the poor. A Reppstiorar's emoluments are excessively small, and his office a very inferior one.

The Landfogued of Iceland is the treasurer of the island, and to that office the one of Byefogued is generally annexed, which is the master of the police in the town of Reikevig.

The court of criminal and civil judicature consists of a judge* and two assessors (or inferior judges) with a secretary. All sentences must be signed by the Stiftsamptman, and an appeal lies from this court to the supreme court of judicature at Copenhagen†. Iceland knows of no trials by jury; for the judge and assessors act both as jurors and judges. Besides this superior court or althing,

* The present chief judge or justitiarius is the learned Mr. Stephensen, whose name so often occurs in the course of the Journal. He is generally called by his Danish title of Etatsroed (Counsellor of State).

† Of late years, in consequence of the difficulty of communication between the parent country and Iceland, supreme power and authority in the courts of judicature have been given to the governor, in conjunction with the chief judge and assessors. This, however, is understood to be only a temporary arrangement.

which has its sittings six times in the year at Reikevig (whither it has been removed only within these few years from Thingevalle), there are annual provincial courts held in the different syssels, and extraordinary ones are occasionally appointed by the Amptmen.

The punishments for capital offences are at present the same as those in Denmark, and the criminal is not hanged but beheaded. It is a fact, however, that of late years, no Icelander has been found who would undertake the office of executioner, so that it has been necessary for the very few who have been sentenced to suffer death, to be conveyed to Norway, there to receive the punishment due to their crimes. The common mode of punishing offences of a less heinous kind, is either whipping, or close confinement and hard labor in the tughthuus, or house of correction, for certain years, or for life.

Of the revenues accruing to the parent state, I am not capable of speaking with any degree of certainty. "Some of them arise from taxes on property, founded upon an estimate which is annually made, under the superintendance of the Reppstiorars of the several individuals in each parish. This estimate is conducted in a somewhat singular way; its basis being a very ancient regulation of property, according to the number of ells of wadmal, the cloth of native manufacture, which each individual possessed, or was enabled to manufacture in the course of the year. The term *hundred,* which was formerly a division derived from the number of ells, is now applied to other descriptions of property. An Icelander is reckoned possessor of an hundred, when he has two horses, a cow, a certain number of sheep and lambs, a fishing-boat, furnished with nets and lines, and forty rix-dollars in specie; and it is by this ratio that the amount of all possessions is ascertained, and the tributes levied upon them. One of the tributes, called the *Tuind's,* requires from every person, possessing more than five hundreds, the annual payment of twelve fish, or an equivalent amounting to twenty-seven skillings, or somewhat more than a shilling of English money. This tax increases in an uniform ratio with the increase of property; and its produce is allotted in equal portions to the public revenue, to the priests, to the churches, and to the maintenance of the poor. Another tribute, called the *Skattur*, consisted, in former times, of twenty ells of wadmal, but is now commuted to money, at the rate of four skillings and an half per ell. It is paid to the public revenue by the owners of farms, and by all, whose property, estimated in hundreds, exceeds the number of individuals composing their families. A third tax, called the *Olaf-tallur*, is paid either in fish or money; likewise in proportion to the property of each individual* " Besides, however, what arises from the taxes imposed upon the inhabitants, the king receives a certain sum for the rental of such farms as are his private property. Land in Iceland comes under three divisions: such as belongs to the king, to the church, and to the peasants themselves. It would be interesting to ascertain, were it possible, the present proportion of each, but to do this with any kind of accuracy is impracticable, from the

* *Dr. Holland, in Sir G. Mackenzie's Travels in Iceland,* p. 323.

various changes that have taken place. The Icelandic Villarium is here our only guide, and from this is extracted the following statement, in applying which to the present time, it must be observed, that, from subsequent sales, the quantity of farms in the possession of the occupiers has been materially increased, and the regal and ecclesiastical estates proportionably diminished.

	No. of Farms.
To the King	718
To the Bishop's see of Skalholt	304
To the Bishop's see at Holum .	345
Church Glebe	640
Glebe of Clergy	140
Glebe of superannuated Clergy ..	45
For maintaining the Poor	16
For maintaining the Hospitals..	4
To Farmers	1847
Total number of Farms ..	4059

The exact expenditure of the island, which, in the present state of affairs, considerably exceeds the amount of the revenues, is more easily ascertained; but, previously to mentioning the particulars of it, it will be necessary to give some little account of the persons holding offices, who have not yet been noticed, but whose expences are defrayed by government, or, what is the same thing, paid from funds established for the purpose, which are under the superintendence of goverment. The salaries of the different masters of the small school at Bessestedr, the only one in the island maintained at the public expence, together with the allowance for the support of the boys, amount to three thousand two hundred and fifty-three rix-dollars.

It is greatly to be lamented that there are no hospitals throughout Iceland of any sort; that which formerly existed at Guvernæs having been dissolved, from being considered too burthensome an institution, and the poor wretches sent to their respective homes, where those deemed incurable are allowed a small pittance for their maintenance, which does not altogether exceed the sum of sixty-four rix-dollars per annum. There is consequently no place of reception for the sick, and, what aggravates the evil is, that there are but six medical men in the whole island, and these necessarily resident at such a distance from the greater number of the inhabitants, that they are comparatively of little service: their salaries are besides extremely small. An apothecary is commissioned to distribute gratis a certain quantity of medicine annually, for which, independent of his pay, he is allowed three hundred and fifty rix-dollars. To judge from all this, it might be concluded that Iceland is singularly salubrious, but, on the contrary, in no country is medical attendance more necessary than here, where the greater part of the inhabitants are afflicted with the most inveterate cutaneous complaints, for which their extreme ignorance and the want of medicines render them incapable of applying either remedy or palliative. The sick and the lame are seen crawling about in almost every part of the island, presenting the most pitiable objects of distress and misery. Nor is more care taken of the females, or of providing for the safety of the coming generation; as, though twenty midwives are provided by government, they are grossly ignorant, and the pains taken to remove their ignorance are so applied as to be almost wholly nugatory. One is sent from Copenhagen for the purpose of giving the necessary instructions to the rest; but her salary of one hundred dollars per annum is too small to enable her to take long journies, or to do any effectual good. The other nineteen receive altogether only one hundred rix-dollars per annum.

I must not omit, in the small list of useful officers in the pay of government, to mention two Danish lieutenants, who are engaged with respectable salaries in the survey of the whole island; and, to judge from one or two specimens of their plans that have come under my observation, they are well capable of undertaking this important task.

The annual expences of Iceland, which are paid by government from various funds established in Copenhagen, will be at once seen by the following accounts. It will be, however, necessary to observe, that $2\frac{1}{3}$ per cent. is deducted by government from all salaries paid to officers and others, unless the contrary is permitted by express order. What is called *extra deduction* in the accounts, seems to be a kind of imposition practiced on some particular persons, since it is not exacted from all alike. Another deduction is also made for *rank-tax*, unless the officers are exempted from it by special permission; and such is the case throughout the whole of his Danish Majesty's dominions: all are obliged to pay a tax in proportion to the rank they maintain; whether this rank is obtained by the employments they hold in the state, or whether it is a mere title.

The current money of the country is chiefly Danish bank-notes of ninety-six skillings value each. One skilling is equal to a halfpenny English, sixteen skillings constitute one mark Danish, and six marks Danish one rix-dollar. These bank-notes are, however, distinguished from those current in Denmark, by having a few Icelandic words printed on the back, specifying their value. The only specie to be met with consists of these skillings, penny, twopenny, and fourpenny pieces of an adulterated silver: all other silver and gold coin, which used to be seen in abundance, is now almost unknown.

A rix-dollar, as just observed, should be equal to four shillings English, and such was the case, or very nearly so, before the breaking out of the war between the two countries, but, at present, on account of the low course of exchange, it is not more than equal to one-third of that value. It will be seen that the regular expenditure of the island is nearly twenty thousand rix-dollars or £4,000 sterling; other occasional expences, however, make it amount to nearly £6,000, that is, thirty thousand rix-dollars. These extra expences are supplied by the King of Denmark in bank-notes, which he annually remits to the island.

Account of Salaries and Pensions paid yearly in Iceland by the Landfogued, Frydensberg.

SALARIES.

From the Jordebog's Casse

			Rdr.	Sk.	Rdr.	Sk.
The Stiftsamptman's regular pay . . .			1200	0		
Deduction	28	0				
Rank-tax deducted	70	0	98	0		
			1102	0		
Augmentation of pay.			300	0		
Total salary with deductions and additional pay					1402	0

			Rd.	Sk.	Rdr.	Sk.
The Amptman over the Western Ampt.						
Regular pay			1000	0		
Deduction	23	32				
Rank-tax deducted	40	0	63	32		
Total sum					936	64
The Amptman over the Northern and Eastern Ampt (independent of the revenue of Mödre valle Cloister, which is paid not in money but in kind). Regular pay			695	4		
Deduction of 2½ per cent. and rank-tax			40	0		
Total sum					655	4
The Chief Judge. Regular pay			900	0		
Deduction of 2½ per cent. and rank-tax			45	0		
			855	0		
Augmentation of pay			300	0		
Total sum					1155	0
First Assessor in the High Court of Judicature.						
Full pay			700	0		
Deduction	16	32				
Rank-tax deducted	15	0	31	32		
Total sum					668	64
Second Assessor in the same court.						
Full pay			500	0		
Deduction	11	64				
Rank-tax deducted	16	0	27	64		
Total sum					472	32

			Rdr.	Sk.	Rdr.	Sk.
The Secretary to the same court.						
Full pay			150	0		
Deduction	3	48				
Rank-tax deducted	6	0	9	48		
Total sum					140	48
The Landfogued of Iceland and Byefogued of Reikevig.						
As Landfogued—Full pay			600	0		
As Byefogued—Ditto			300	0		
Deduction from the latter			7	0		
Total sum					893	0
To the Police Officers in Reikevig, each without deductions			150	0	300	0
To the Sysselman of Westmann's Islands, without deductions					57	48
To the Sysselman in Kiöse and Guldbringue Syssels, who is at the same time administrator of the King's estate in the latter syssel.—Full pay, without deduction			33	72		
Augmentation of pay, which is liquidated in the revenues in the two syssels			200	0		
Total sum					233	72
First Surveyor of the Island.						
Pay and emoluments			825	0		
Sum allowed for travelling expences			350	0		
Total sum					1175	0

			Rdr.	Sk.	Rdr.	Sk.
Second Surveyor of the Island.						
Pay and emoluments			930	0		
Sum allowed for travelling expences			350	0		
Sum total					1280	0
The Chief Physician and Surgeon.						
Full pay			600	0		
Deduction	14	0				
Further extra deduction	60	0	74	0		
			526	0		
Allowed in lieu of an assistant yearly			60	0		
Total sum					586	0
The Government Apothecary.						
Full pay			50	0		
Augmentation of pay			80	0		
			130	0		
Deduction			3	4		
			126	92		
Allowed for medicines for the poor			350	0		
Total sum					476	92
To the Midwife, Madam Malanquist, without deductions					100	0
To all the other Midwives on the island, jointly					100	0
The Surgeon in the Southern Ampt has no pay, but is allowed yearly, to indemnify him for lands to which he is entitled, the sum of					12	0

	Rdr.	Sk.	Rdr.	Sk.
The Surgeon of the first district in the Western Ampt			49	77½
The Surgeon in the second district in the Western Ampt	49	77½		
Allowed yearly to indemnify him for lands	8	11		
Total sum			57	88½
The Surgeon in the Northern Ampt ..			49	77½
The Surgeon in the Eastern Ampt....			49	77½
Certain sums allowed yearly for the augmentation and increase of poor clergymen's salaries			318	0

From the School Funds.

	Rdr.	Sk.	Rdr.	Sk.
Bishop of Iceland—Regular pay	1248	0		
Augmentation of pay, all without deductions	600	0		
Total sum			1848	0
Lecturer on Theology, Bessestedr School			600	0
The Priest of the Church at Reikevig	24	0		
Deduction	1	42		
Total sum			22	54
To the Stiftprovst, Dean of all Iceland			16	0

To the Inspector, or Steward, of Bessestedr School, who undertakes to provide the

	Rdr.	Sk.	Rdr.	Sk.
scholars with necessaries, and to see them regularly distributed.				
Salary	30	0		
For fuel	50	0		
	80	0		
Deduction of 6 per cent	4	77		
	75	19		
Receives yearly as a gift	150	0		
	225	19		
Deduction	3	48		
	221	67		
Receives annually, for 24 scholars, 60 rix-dollars each, for their maintenance.	1440	0		
Total sum			1661	67
To two Teachers in the School of Bessestedr, each per annum	300	0	600	0

PENSIONS

Paid out of the Jordebog's Casse, or from Funds not mentioned or properly regulated.

To the Sysselman in Vesterskaptar-fel Syssel (as liquidated in revenues from Tykebag Cloister)			30	0

	Rdr.	Sk.	Rdr.	Sk.
To the Sysselman in Skagefiord's Syssel	30	0		
Ditto in Kiöse Syssel	30	0		
Ditto in Barderstrand Syssel	60	0		
Ditto in Myre Syssel	30	0		
Ditto in Norder Mule Syssel	30	0		
Total sum			210	0
To Surgeon Backmann			20	0

Paid from the Skatkammer Casse (Treasury Chest.)

	Rdr.	Sk.	Rdr.	Sk.
To the former Stiftsamptman, Olav Stephensen	800	0		
Deduction	18	64	781	32
To Sysselman Snorrasen's Widow			20	0

Paid out of the Post Casse.

	Rdr.	Sk.
Allowed for the augmentation or amendment of the income of Clergymen's Widows	300	0
To John Olafsen's Widow	40	0
To Magnus Olafsen's Widow	50	0
To Snorre Biörnsen's Widow	30	0
To Surgeon Petersen's Widow	16	0
To Landfogued Skulesen's Widow	25	0
To Sysselman Snorresen's Widow	16	0
Ditto Arnersen's ditto	30	0
Ditto Thomassen's ditto	23	0
Ditto Einersen's ditto	15	0

	Rdr.	Sk.
To Pastor emeritus Gudmun Poulsen, in Kaloholt	20	0
To Surgeon Halgrim Backmann	20	0
To Surgeon Brynjole Petersen	60	0
To Nicolaysen's Widow	20	0

Paid from the Rentekammer's Poor-Box.

To former Under-Assistant Jon Olsen	20	0

Paid from the Danish War-Hospital Funds.

To Invalid Jon Einersen	12	0

Paid from the Icelandic and Finmarkish Company Funds.

To the former Under-Assistant to the Company	50	0

Paid from the former Guvernæs Hospital Funds.

To Thorkel Gissursen, Biarne Gissursen, and Gudrun Snorredatter, all in Kiöse Syssel	26	0
To Gudmun Thorlaksen of Asum in Hunevald Syssel	20	0
To Ingwald Einardatter in Arnæs Syssel	6	0
To Oddni Kehildsdatter in Dale Syssel	6	0
To Gunhild Jonsdatter in Guldbringue Syssel.....................	6	0
To Olav Jonsen in Havnfiord	6	0

Paid from the School Funds.

	Rdr.	Sk.	Rdr.	Sk.
To former Corrector Paul Jacobsen ..	100	0		
Deduction 6 per cent	6	0		
Total sum			94	0
To Einar Biarnesen in Arnæs Syssel ..			5	0
To Bishop Stephensen's Widow, at Holum	120	0		
Deduction 2 77				
Extra deduction 2 38	5	19		
Total sum			114	77
Former Rector Paul Hialmersen	150	0		
Deduction $2\frac{1}{2}$ per cent. and extra deduction	8	0		
Total sum			142	0
To the poor, in the service of the King when he had the whole trade of the island			287	0

Note—The sum to these is often liquidated out of the royal taxes and paid by the Sysselman of each district.

Paid from the Funds established to meet the expences of the Post.

	Rdr.	Sk.
Former Postman, Vigfus Jonsen......	3	0
Former Postman, Sunner Ravsen	5	0
Annual expences attending the Post ..	300	0

Total Amount of the yearly Expenditure of the Island of Iceland, in Salaries and Pensions, as paid by the Landfogued.

	Rdr.	Sk.
Salaries paid out of the Jordebog's Casse, that is of the Funds established out of Royal or Episcopal Estates in Iceland. .	11169	73
Salaries paid out of the School Funds, to the Clergy and Teachers	4743	73
Pensions paid out of the Skatkammer Casse, or Treasury	801	32
Pensions paid out of the Post Casse, or Post Funds	672	0
Pensions paid out of the Rentekammers, or Chamber of Rents Poor-box	20	0
Pensions paid out of the Danish War-Hospital Funds	12	0
Pensions paid out of the Icelandic or Finmarkish Company's Funds	50	0
Pensions paid out of the former Guvernæs Hospital Funds	64	0
Pensions paid out of the School Funds	642	77
Pensions paid out of the Funds established to pay the expences of the Post in the Country	308	0
Total sum	18713	63

Having thus, in a very cursory manner, noticed a few of the most important circumstances, connected with the civil and political affairs of the country, I shall proceed to some brief remarks on the religious history of the ancient northern nations, and of Iceland in particular; in doing which, I shall make ample use of the valuable information contained in the "*Northern Antiquities*" of M. Mallett.

The religion of the north, in its greatest purity, taught the existence of a supreme God, "the author" according to the Icelandic Mythology, "of every thing that existeth; the eternal, the ancient, the living and awful being, the searcher into concealed things, the being that never changeth"; to whom, also, was attributed "a boundless knowledge and an incorruptible justice." From him sprung (as it were emanations of his divinity) an infinite number of subaltern deities and genii, of which every part of the visible world, was the seat and temple. These intelligencies were not contented barely to reside in each part of nature, but they directed its operations, and it was the organ or instrument of their love or liberality to mankind. Each element was under the guidance of some being peculiar to it. The earth, the water, the fire, the air, the sun, the moon, and the stars, had each their respective divinity; and to serve these several gods with sacrifices, to be brave and intrepid themselves, and to do no wrong to others, were the moral obligations inculcated upon mortals by this religion. To these duties was added the belief in a future state, where cruel tortures were reserved for those who despised the three fundamental precepts of morality, and joys without number for the religious, just, and valiant. This appears to have been the state of religion among the Scandinavians, till towards the period of the fall of the Roman empire, when, in consequence of the arrival of Odin in the north; it began to lose much of its original purity. The people became weary of its simplicity, and associated to the supreme God many of those genii or inferior divinities, who had always been subservient to him; and even the supreme being himself, whom they called by the name of Odin, they divested of a portion of his omnipotence, and looked upon him as little more than the god of war, in which character he is called in the Edda, "the terrible and severe god, the father of slaughter, the god that carrieth desolation and fire, the active and roaring deity, he who giveth victory, and reviveth courage in the conflict; who nameth those that are to be slain." Such as were most brave in battle, and as died fighting, were received by him in his palace Valhala: thus, when Regner Lodbrog* was at the point of death, far from uttering complaints, he burst out into an exclamation of rapture; "We are cut to pieces with swords: but this fills me with joy, when I think of the feast that is preparing for me in Odin's palace. Quickly, quickly seated in the splendid habitation of the gods, we shall drink beer† out of the skulls of our enemies. A brave man fears not to die. I shall utter no timorous words as I enter the hall of Odin". Next to Odin, Freya, his wife, was considered the principal

* See translations from the Icelandic, entitled *Five Pieces of Runic Poetry*. p. 27.

† Odin alone drank wine.

deity, who appears to have been the Venus in the northern Mythology; and next to her was Thor, whose authority extended over the winds and seasons, and particularly over thunder and lightning. He is called in the Edda, the most valiant of the sons of Odin. These three deities composed the supreme counsel of the gods, and were the principal objects of the worship of the Scandinavians, who, nevertheless, were not all agreed about the preference which was due to each of them in particular: thus the Danes paid the highest honors to Odin, and the Swedes to Freya, while the natives of Iceland bestowed them upon Thor. Twelve other gods (inferior deities) and as many goddesses are besides enumerated in the Edda. Odin was believed to be the creator of heaven and earth. The ideas upon this head, as handed down to us by the Icelandic Mythology, cannot be better expressed than in the language of the Voluspa. "In the day-spring of the ages," says the poet, "there was neither sea nor shore, nor refreshing breezes. There was neither earth below nor heaven above to be distinguished. The whole was only one vast abyss, without herb, and without seeds.

The sun had then no palace; the stars knew not their dwelling-places; the moon was ignorant of her power. After this there was a luminous, burning, flaming world towards the south, and from this world flowed out incessantly into the abyss, that lay towards the north, torrents of sparkling fire, which, in proportion as they removed far away from their source, congealed in their falling into the void, and so filled it with scum and ice. Thus, was the abyss, by little and little, filled quite full: but there remained within it a light and immoveable air, and thence exhaled icy vapors. Then a warm breath coming from the south melted those vapors, and formed of them living drops, whence was born the giant Ymer. It is reported that, while he slept, an extraordinary sweat, under his armpits, produced a male and female, whence is sprung the race of giants; a race evil and corrupt, as well as Ymer their author. Another race was brought forth, which formed alliances with that of the giant Ymer: this was called the family of Bor, so named from the first of that family, who was the father of Odin. The sons of Bor slew the giant Ymer, and the blood ran from his wounds in such abundance, that it caused a general inundation, wherein perished all the giants, except only one, who, saving himself in a bark, escaped with all his family. Then a new world was formed. The sons of Bor, or the gods, dragged the body of the giant into the abyss, and of it made the earth: the sea and rivers were composed of his blood; the earth of his flesh; the great mountains of his bones; the rocks of his teeth and of the splinters of his smashed bones. Of his skull they formed the vault of heaven, which is supported by four dwarfs, named south, north, east, and west. They fixed there tapers to enlighten it, and assigned to other fires certain spaces which they were to run through, some of them in heaven, others under the heaven: the days were distinguished and the years were numbered. They made the earth round, and surrounded it with the deep ocean, upon the banks of which they placed the giants. One day it chanced, as the sons of Bor, or the gods, were taking a walk, they found two pieces of wood floating upon the water; these they took, and out of them made a man and a woman. The eldest of the gods gave them life and souls; the second, motion and knowledge; the third, the gift of speech, hearing, and sight, to which he added beauty and raiment. From this man and this woman, named Askus and Embla, is descended the race of men who are permitted to inhabit the earth." It is easy, as M. Mallet observes, to trace out, in this narration, vestiges of an ancient and general tradition, of which every sect of paganism hath altered, adorned, or suppressed many circumstances, according to its own fancy, and which is now only to be found entire in the books of Moses.

Superstition held great sway over the minds of the pagans, and magicians and sorcerers abounded *.

Upon the subject of the final dissolution of the world, and the notions entertained by

* A long and interesting history of the different kinds of magic among the Icelanders, both during the continuance of paganism, and for a considerable period after, may be seen in the *Voyage en Islande*, *v.* III. *p.* 78 *and seq.*

these people of a future state, I shall again have recourse to the Edda and the Voluspa, as translated in the *Northern Antiquities*.

"There will come a time," it is declared, "a barbarous age, an age of the sword, when iniquity shall infest the earth, when brothers shall stain themselves with brothers' blood, when sons shall be the murderers of their fathers, and fathers of their sons, when incest and adultery shall be common, when no man shall spare his friend. Immediately shall succeed a desolating winter, the snow shall fall from the four corners of the world, the winds shall blow with fury, the whole earth shall be hard bound in ice. Three such winters shall pass away, without being softened by one summer. Then shall succeed astonishing prodigies: then shall the monsters break their chains and escape: the great dragon shall roll himself in the ocean, and with his motions the earth shall be overflowed: the earth shall be shaken, the trees shall be torn up by their roots: the rocks shall be dashed against each other. The wolf Fenris, broke loose from his chains, shall open his enormous mouth, which reaches from heaven to earth; the fire shall flash out from his eyes and nostrils; he shall devour the sun; and the dragon, who follows him, shall vomit forth, upon the waters and in the air, great torrents of venom. In this confusion, the stars shall fly from their places, the heavens shall be cleft asunder, and the army of evil genii and giants, conducted by Sortur (the black), and followed by Loke, shall break in to attack the gods. But Heimdal, the door-keeper of the gods, rises up; he sounds his clanging trumpet; the gods awake and assemble; the great ash tree shakes its branches; heaven and earth are full of horror and affright. The gods fly to arms; the heroes place themselves in battle array. Odin appears armed in his golden casque and his resplendent cuirass: his vast scymitar in his hands. He attacks the wolf Fenris, by whom he is devoured, and his antagonist perishes at the same instant. Thor is suffocated in the floods of venom, which the dragon breathes forth as he expires. Loke and Heimdal mutually kill each other. The fire consumes every thing, and the flame reaches up to heaven. But, presently after, a new earth springs forth from the bosom of the waves, adorned with green meadows: the fields there bring forth without culture; calamities are there unknown; and a palace is there raised more shining than the sun, all covered with gold. This is the place that the just will inhabit, where they will enjoy delights for evermore. Then the powerful, the valiant, he who governs all things, comes forth from his lofty abodes to render divine justice. He pronounces decrees. He establishes the sacred destinies which shall endure for ever. There is also an abode remote from the sun, the gates of which face the north, and poison rains there through a thousand openings. Through this place, which is all composed of the carcasses of serpents, run certain torrents, in which are plunged perjurers, assassins, and those who seduce married women. A black-winged dragon flies incessantly around, and devours the bodies of the wretched who are there imprisoned."

From this slight sketch it appears that the northern nations believed in the immortality of the soul, as well as in the existence of a future state of happiness and misery; and, moreover, that there were two abodes destined for each of these states. To the former belonged Valhala, the palace of Odin, where all were admitted who had died* a violent death, from the time of the creation of the world to the period of the universal dissolution of nature, and Gimle, or the palace covered with gold, where the just were to enjoy delights for ever. On the

* "The heroes," says the Edda, "who are received into the palace of Odin, have every day the pleasure of arming themselves, of passing in review, of ranging themselves in order of battle, and of cutting one another in pieces; but, as soon as the hour of repast approaches, they return on horseback, all safe and sound, back to the hall of Odin, and fall to eating and drinking. Though the number of them cannot be counted, the flesh of the boar, Servimner, is sufficient for them all; every day it is served up at table, and every day it is renewed entire. Their beverage is beer and mead; one single goat, whose milk is excellent mead, furnishes enough of that liquor to intoxicate all the heroes: their cups are the skulls of enemies they have slain. Odin alone, who sits at a table by himself, drinks wine for his entire liquor. A crowd of virgins wait upon the heroes at table, and fill their cups as fast as they empty them." *Northern Antiquities. v.* I. *p.* 120, and *Edda Iceland. Mythol.* 31, 33, 34, and 35.

other hand, Niflheim (from the Icelandic *Nifl*, evil, and *Heim*, home) is the first of the abodes of misery, which was only to exist till the renovation of the world; while the second, named Nastrond (the shore of the dead), was to continue for ever.

Such were the doctrines taught by the religion of the early inhabitants of Iceland, if we may give credit to the histories of their mythology that are handed down to us. I shall now say a few words of their places of worship and of their religious ceremonies. The former, called Hoff, we are told by Arngrim Jonas, were of great dimensions, and, for such a country, of magnificent structure. One of these, situated in the prefecture of Watzdal, in the northern part of the island, is spoken of as being one hundred and twenty feet in length, and another, at Kialarnes in the south, sixty feet long. To each temple was annexed a small building or chapel, which was esteemed the most sacred place; for here the idols were kept standing upon a pediment, and around them were arranged the beasts that were to be sacrificed. The chief of these idols was Thor, who was placed in the centre of the minor deities*. Immediately before the gods, also, stood an altar, cased with iron, lest it should be destroyed by the continual fires. Here also stood a large brazen vessel, in which was poured the blood of the victims; and here, too, were the purifying instruments (lustrica) and the brushes for sprinkling the congregation with blood, together with a ring of silver, or of brass, twenty inches long, which was held by those who made oath†. The victims

* What these deities were, or what was their number, does not seem to be rightly known. Arngrim Jonas mentions three, besides those who were invoked during the time that the rite was celebrated, which was always performed when a person made oath upon the most solemn occasion. "In veteri tamen juramenti formulâ, tres præter Thorum nomine notantur: *Freyr*, *Niordur*, *As*. Quorum tertium, nempe *As*, existimo esse *Odinum* illum famosum, inter divos ethnicos non postremum habitum dictum *As*, quod is Asianorum huc in septentrionem migrantium princeps fuerit: singulariter nempe *As*, at multitudinis numero *Aesar* vel *Aeser* dici cœperunt." *Tractatus de Islandiâ*, p. 430.

† When any person was suspected of having spoken falsely in an affair of importance, he was put to his oath, and then his veracity was determined by making him stand under an oblong piece of turf, placed in

slaughtered were generally sheep and oxen, and those parts which were not consumed in the sacrifice, were considered as belonging to the officiating priest. These animals, however, were not at all times looked upon as a sufficient atonement or propitiatory offering, whence it happened that, in case of any extraordinary crime, calling for extraordinary vengeance from the gods, the altar flowed with the blood of human victims, and, at Kialarnes, a deep pit or well was formed near the chapel, into which these unhappy

such a manner that it should form, over him, an arch, with its extremities touching the ground: if it supported itself without breaking, the man was declared worthy of belief, if otherwise, he was condemned. But when two or more persons were about to join in a covenant, the arched piece of turf was supported by a lance, and those engaged in the treaty placed themselves beneath it, where each with a sword drew blood from himself, and mingled it with that of his companions, as a sign of mutual faith. By this ceremony the most powerful compact was sworn and ratified; and, besides the mutual aid which, by this act, they were obliged to afford each other during life, if any were slain, the survivors, how many soever there might be, were bound, in the most sacred manner, to revenge his death by all the means in their power. *Arngrim Jonas, Tract. de Islandiâ.*

wretches were cast, and which thence bore the name of Blotkelda. So, likewise, in the province of Thornes-thing, there was a similar excavation, in which were confined those who were destined to be offered as a sacrifice to the gods, and who were thence selected and killed upon a large stone, "cujus rei indignitatem," says Arngrim Jonas, "saxum illud fertur colore sanguinolento nullo imbre abluto multis post seculis retulisse." The same learned author, however, anxious in some manner to extenuate, if not to justify, the atrocities of his countrymen, asserts that human sacrifices were more common in other countries of the north than in Iceland, in which, he assures us, they took place only in two provinces, and even there all the inhabitants did not join in them. Hiorleif, the companion of Ingulf, renounced altogether the worship of idols. Helgo, whose surname was Biole, a native of Kialarnes, a man of high rank, and descended from the Norwegian barons, did not countenance the religion of the pagans, but offered his protection to a christian exile from Iceland, whom he permitted to build a temple, and to dedicate it to St. Columbus at Esinberg. A young man, also, called Buo, living in the same province, destroyed, by fire, the gods belonging to the temple held in the highest veneration, the same in which the human victims were sacrificed. The name of Thorcillus, too, who flourished about the year 900, and was at that period the Logman or chief magistrate in the island, deserves to be commemorated as superior to the superstitions of his age and country. He, finding himself drawing towards the close of his existence, gave orders that he should be taken into the air, and that his face should be turned towards the sun; when, having remained for some moments in a kind of ecstacy, he expired, recommending his soul to Him among the gods, who had created the sun and the stars.

It was not till A. D. 974, in the reign of Olaf I. of Norway, that an attempt was made at introducing the christian religion. Frederic, a Saxon bishop, arrived in 981, and preached the gospel with such success, that in 984 churches were built, and many persons received baptism. Still, however, no material progress was made; for Bishop

Thangbrandt and Stefr Thorgilsen, who were sent from Germany in the year 997, were received with stones, and they and their religion abused with the keenest invectives by the poets of that day. Through the exertions of these, however, and other missionaries, the light of christianity began more and more to shed its lustre upon the minds of the people, so that, on the arrival of Gissur and Hjatle in the year 1000, the whole island became converted, without bloodshed, though not without opposition; and it was agreed, at a general assembly of the inhabitants, that the worship of idols should be abandoned, and the religion of our blessed Saviour * embraced in its stead.

* It appears that, at this time, the rite of baptism was celebrated at one of the hot-springs in the neighborhood of the famous Snorralaug, noticed at p. 311 of this volume; for Eggert Olafsen, after speaking of the Nyrdre-Reykia-dal, says, "Huic collateralis Sydre-Reykiadalur, vallis fontibus fervidis abundans: hìc est Kros-laug, balneæ limpidæ et salubres, in quibus anno Christi millessimo, Islandiæ occidentalis Incolæ, abominantes aquam frigidam, sacro baptismate abluti sunt, unde balneis crucis nomen erat impositum." *Enarrationes Historicæ de naturâ et constitutione Islandiæ.* p. 31.

In 1050, it was farther decreed, in a solemn assembly, that the temporal or politic law, which was introduced from Norway by one Ulfliot*, in the year 928, should every where give place to the canon or divine law.

After this period monks and convents began to abound in the island, and the people paid a yearly tribute to the Roman see of ten ells of wadmal for each family.

In the year 1056 the Icelanders received the first of their bishops, Islief, who was consecrated to the see of Skalholt, and a second was instituted to that of Holum, in 1107. Both were originally under the jurisdiction of the Archbishop of Bremen and Hamburgh; but, in 1103 or 1104, they became subordinate to Azerus, the first Archbishop of Lund in Scania, and in 1152 to the Bishop of Drontheim.

* *Ulfliots Laug* (le code d'*Ulfliot)* fut le premier code de l'Islande, et en grande partie incorporé dans celui d'aujourdhui, nommé *Graagaasen*; son auteur fut le prévôt *Ulfliot* qui demeurait dans ce quartier, dans le canton de *Lon*; il fit accepter ce code en 928, et le tribunal supérieur, nommé *Althing*, fut établi peu après. *Voyage en Islande.* IV. p. 44.

The Lutheran religion was introduced by King Christian III. in the year 1540, but the zeal, with which the bishops opposed this new sect, prevented it from gaining ground till 1551; when the last and most earnest supporter of popish superstitions, Jon Areseni* was beheaded by order of the King's Lieutenant. Soon after this period all the inhabitants embraced the Lutheran faith.

Iceland at present has only one bishop; for, in the year 1785, the King of Denmark ordered that all the estates belonging to the see of Skalholt should be sold, and the money deposited in the funds called Jordebog's Casse. The episcopal see was removed to Reikevig, and a certain yearly salary granted to the bishop in lieu of his former privileges. So also were the estates belonging to Holum sold in the year 1801; the money secured in the same funds; and the two dioceses incorporated into one. Among the Danish clergy there is no metropolitan or archbishop, but each bishop has the full

* Arngrim Jonæ Comment. de Islandiâ.

power over his own stiftsampt, without being subject to any other ecclesiastical jurisdiction, though he is amenable to the civil powers. In like manner the Bishop of Iceland is independent of all other bishops.

The next officer in the church is the Stiftprovst of all Iceland, which is somewhat analogous to a dean in England. The present Stiftprovst's name is Magnusen.

The Provsts are inferior officers of the diocese, who have the care and superintendence of ecclesiastical affairs in their own provstie; for the diocese is divided into nineteen such provsties, and about one hundred and eighty-four church livings.

The priests receive their income from the lands that are annexed to each church and from tythes; besides which, there are three hundred and eighteen rix-dollars and seventy-two skillings allowed per annum for the amendment of such livings as are very small, and three hundred more for the augmentation of pensions to poor clergymens' widows. Their salaries are various; a few exceed a hundred rix-dollars per annum, but the greater number do not produce an income of more than thirty or forty rix-dollars, and some do not exceed twelve, ten, or even five. It must be remembered, however, that most of the clergy occupy little farms, and this alone makes the condition of the greater part of them tolerable.

To give a more correct idea of the revenues of the clergy of Iceland, not only of the regular salaries which they receive from the crown, but also of such pensions as are appropriated to superannuated and poor priests and widows, I subjoin the following table of expenditure; only premising, that the number of clergymen is not to be estimated by that of the livings here mentioned; for, curates included, I think they amount to between three and four hundred.

			Rdr.	Sk.
1	Osterskaptefields Provstie	5 Livings	113	20
2	Vesterskaptefields Provstie....	7 Ditto	164	40
3	Rangervalle and Westmannöe Provstie	13 Ditto	736	72
4	Arnæs Provstie..............	16 Ditto	436	34
5	Guldbringue and Kiöse Provstie	9 Ditto	349	43
6	Borgafiord Provstie..........	6 Ditto	216	28
	Carried forward		2016	45

			Rdr.	Sk.
	Brought forward		2016	45
7	Myre Provstie	7 Livings	338	18
8	Snoefieldnes Provstie	7 Ditto	455	7
9	Dale Provstie	6 Ditto	281	18
10	Barderstrands Provstie	8 Ditto	291	72
11	Isefiords Vester Provstie	6 Ditto	215	80
12	Isefiords Norder Provstie	7 Ditto	188	41
13	Strande Provstie............	4 Ditto	103	46
14	Hunevands Provstie	15 Ditto	453	31
15	Skagefiords Provstie	14 Ditto	403	50
16	Oefiords Provstie............	15 Ditto	434	16
17	Norder Provstie	17 Ditto	668	15
18	Norder Mule Provstie........	10 Ditto	255	40
19	Syder Mule Provstie	12 Ditto	359	90
	Total		6464	89

	Rdr.	Sk.
The amount of the revenues of the church-lands and tythes is therefore	6464	89
To this may be added,		
The Bishop's annual salary	1848	0
The Stiftprovsts annual salary	16	0
Salary to the Priest of Reikevig church	22	54
Pension to Bishop Stephensen's Widow	117	19
Pension to Pastor emeritus Bergsen	20	0
Pension to Pastor emeritus Tholevsen......	30	0
Total sum	8518	66
The sum for the augmentation and amendment of poor clergymen's livings and widow's pensions is	618	72
Thus the grand total of the expenditure of the church amounts to	9137	42

The Icelandic language is the most ancient, and most pure, of all the Gothic and Teutonic dialects. It has been called the Cimbric, from its having been the one which chiefly prevailed among those tribes who inhabited the Cimbrica Chersonesus, and, under this name, it is considered by the learned Dr. Hickes *, as the parent of the Swedish, Danish, and Norwegian languages, in the same manner as the Anglo-saxon is of the English, of the Friezland, and of the Lowland Dutch, and the Francic of the German language. All of them proceed from the same original stock †. That the Icelandic has retained its original purity to such a degree, that an Icelander of the nineteenth century can read, with ease, the oldest manuscripts

* In his *Linguarum vet. Septentrionalium Thesaurus Grammatico-criticus et Archæologicus.*

† "To the old original mother-tongue it has been usual, after Verstegan, to give the name of Teutonic, not so much from the Teutones or Teutoni, who inhabited the Danish islands, and were brethren to the Cimbri, as from its being the ancient Tuytsh, the language of Tuisto and his votaries; the great Father and Deity of the German tribes." *Northern Antiquities*, I. p. xl.

of his country, seems to be attributable to the little intercourse which this island has had with foreign nations, and to the small alteration that has taken place in the state of civilization of its inhabitants; few or no strangers having settled among them, who might corrupt their language by a mixture of their own; and few luxuries having been introduced, which might give rise to new wants, and consequently render necessary new terms to express them. What is spoken by the people of the coast is, however, in some degree, altered by the visits of foreigners; and in the immediate vicinity of the ports will be heard a number of words, which have been adopted from the Danes and Norwegians.

A specimen of the modern Icelandic will be found at page 295, of the second volume of this work, in a poem, written by one of the first native scholars of the present day; but, in order to shew how trifling is the change that has taken place in the language, between the years 1585 and 1746, I shall here subjoin a transcript of the Lord's Prayer, as it was written at each of those periods.

ICELANDIC LORD'S PRAYER IN 1585.

Fader vor thu sem ert a himnum. Helgist nafn thitt. Tilkome thitt riike. Verde thinn vilie so a jordu sem a himne. Gief oss i dag vort dagligt braud. Og fyrerlat oss vorar skullder, so em vier fyrerlautum vorum skulldunautum. Og inleid oss ecki i freistne. Helldr frelsa thu oss af illu, thuiat thitt er riikit, maatur og dyrd um allder allda. Amen.

ICELANDIC LORD'S PRAYER IN 1746.

Fader vor thu sem ert a himnum. Helgesst thitt nafn. Tilkomme thitt rike. Verde thin vilie, so a jordu sem a himne. Gief thu oss i dag vort daglegt braud. Og fyrergief oss vorar skullder, so sem vier fyrer-giefum vorum skulldnautum. Og innleid oss ecke i freistne. Helldur frelsa thu oss fra illu; thuiad thitt er riiked og maattur og dyrd um allder allda. Amen.

The Runic* characters, the first in use among the Icelanders, are of very remote antiquity, but of doubtful origin; though the Scandinavians, among whom they seem to have originated, were disposed to attri-

* The word RUNE, Wormius derives either from *Ryn*, a furrow, or *Ren*, a gutter or channel. As these characters were first cut in wood or stone, the resemblance to a furrow or channel, would easily suggest the appellation. *Northern Antiquities*, I. p. 63.

bute the invention of them to Odin. These letters are extremely unlike any that have been in use in other countries, and are only sixteen in number. They were used for the purpose of writing as well as in magical operations*. Many ancient monuments engraven with Runic inscriptions have been found in Iceland, as well as in Denmark and

* In the Havamal, or Sublime Discourse of Odin, it is said "Runic characters destroy the effect of imprecations"; and in Resenus' edition of the Fragments of the Ancient Edda, a little Poem is added, which is called "the Runic Chapter, or the Magic of Odin." In it that conqueror relates the wonders he is able to perform, either by means of these characters, or by the operations of poetry. "Do you know," says he, "how to engrave Runic characters? how to explain them? how to procure them? how to prove their virtue? If I see a man dead, and hanging aloft on a tree, I engrave Runic characters so wonderful, that the man immediately descends and converses with me:" and Angantyr, in the *Incantation of Hervor* (of which we have a translation in the *Five Pieces of Runic Poetry)*, says to Hervor, "Young maid, I say thou art of man-like courage, who dost rove about, by night, among tombs, with a spear engraven with magic spells, with helmet and coat of mail, before our hall:"—these magic spells were Runic characters, engraven on the weapon to prevent their being dulled, or blunted by inchantment.

Norway, and even in England, as mentioned by Hickes; and a vast number of books, written in this character *, still exist in the libraries of the north; but of these, M. Mallet observes that the most ancient appear to have been written about the time that christianity took place in the north, as is judged from several proofs, particularly from the frequent

So, too, in the *Ode on the Descent of Odin*, when this "Father of Magic, having reached the deep abode of death, stops near the tomb of the prophetess and looks towards the north, he engraves Runic characters on her tomb; and he utters mysterious words."

"Right against the eastern gate
By the moss-grown pile he sate;
Where long of yore to sleep was laid
The dust of the prophetic maid.
Facing to the northern clime,
Thrice he traced the Runic rhime;
Thrice pronounced, in accents dread,
The thrilling verse that wakes the dead;
Till, from out the hollow ground,
Slowly breathed a sullen sound."

Gray's Translation of the Descent of Odin.

* Printed characters may be seen in the first volume of *Northern Antiquities*, p. 370; fac similes of some inscriptions, in the Atlas of the *Voyage en Islande*, t. xx, and in the title-page of the *Five Pieces of Runic Poetry*.

intermixture of Roman letters in them. In the year 1000, Isleif founded a school at Skalholt, and soon after four other, when the Roman characters were universally adopted, and the youth instructed in the Latin tongue, divinity, and parts of theoretic philosophy. At this period, also, many Icelanders studied at foreign universities, though others received their education entirely in their own country. Iceland was now in the zenith of her literary glory, and, from the introduction of the christian religion till the year 1264, when the whole island became subject to Norway, she continued one of the few countries in Europe, and the only one in the North, where the sciences were cultivated and held in esteem *. It appears extraordinary, says M. Mallet, to hear a historian of Denmark cite for his authority the writers of Iceland; but this wonder will cease, when the reader is informed, that, from the earliest times, the inhabitants of that island had a particular fondness for history, and that from among them have sprung those poets, who, under the name of *Scalds*, rendered themselves so

* Von Troil.

famous throughout the north for their songs, and for the credit they enjoyed with kings and people. In fact, they have always taken great pains to preserve the remembrance of every remarkable event that happened, not only at home, but among their neighbors, the Norwegians, the Danes, the Swedes, the Scotch, the English, the Greenlanders, &c. The first inhabitants of Iceland carried with them the verses, together with other historical monuments of former times; and the odes of these Icelandic Scalds were continually in every body's mouth; containing, according to Torfæus, the genealogies and exploits of kings, princes, and heroes: and, as the poets did not forget to arrange them according to the order of time, it was not difficult for the Icelandic historians to compose afterwards, from such memoirs, the chronicles they have left us. Indeed *, the poetical and historical works of this country have bid defiance to time. Her ancient chronicles shew what clear notions they had of morality, philosophy, natural history, and astronomy. Her divines read the works

* Von Troil.

of the fathers of the church; and no fewer than two hundred and thirty poets *, some of them known and esteemed at foreign courts, are enumerated in the Skaldatal, an ancient manuscript, in which is preserved a list of those scalds or poets who have distinguished themselves in the three northern kingdoms, from the reign of Regner Lodbrog to that of Valdemer II: that is, from A.D. 750 to 1157. Among them is more than one crowned head, and, what is no less remarkable, the greatest part of them are natives of Iceland. Driven, perhaps, by poverty, some of them were induced to visit foreign courts, and Wormius, in his *Litteratura Danica*, states that Canute had no less than eight Danish, Norwegian, and Icelandic poets, who flourished in his court and enjoyed his friendship. Others doubtless travelled in distant countries for the sake of acquiring knowledge †.

* *Northern Antiquities*, I. p. 391.

† "Præstantes olìm (Islandi)
Relictis patriis oris, Londinum studiosè petebant,
Artium addiscendarum cupidi,
Quas contenta libris eruditio commendat.

Of the ancient Icelandic poems* the Edda† doubtless holds the first rank among

Oxoniæ in Anglico solo
Pedem hoc temporis tractu figere
Imprimis arridebat;
Deinde fortunæ favore suffulti
Solum natale repetebant"

See the Carmen Gratulatorium, v. II. *p.* 280.

* The stile of the ancient poems is very enigmatical and figurative, very remote from the common language, and for that reason grand but tumid; sublime but obscure. If it be the character of poetry to have nothing in common with prose, if the language of the gods ought to be quite different from that of men, if every thing should be expressed by imagery, figures and hyperboles and allegories, the Scandinavians may rank in the highest class of poets. They seldom expressed heaven by any other name than "the skull of the giant Ymer." The rain-bow was called "the bridge of the gods." Gold was "The tears of Freya." Poetry, the "present or the drink of Odin." The earth was either "the spouse of Odin, the flesh of Ymer, the daughter of the night, the vessel which floats on the ages, or the foundation of the air." Herbs and plants were the "hair or the fleece of the earth," &c, *Northern Antiquities*, p. 393 and 395.

† *Edda* is said to be derived from a Gothic word, signifying *Grandmother*, which, in the figurative sense of the old poets, was intended to express an ancient doctrine.

those that have been handed down to us, and the lover of northern antiquities will find an ample store of information upon the subject, in the second volume of M. Mallett's work. It is there stated that there have been two poems of this name, the first and most ancient of which was compiled by Soemund Sigfussen, surnamed the learned, born in Iceland about the year 1057. This author had studied in Germany, and chiefly at Cologne, along with his countryman Are Frode, who distinguished himself by his love for literature. Soemund was one of the first who ventured to commit to writing the ancient religious poetry* which many people still retained by heart. This first collection being too voluminous, Snorro Sturleson, about one hundred and twenty years after, undertook to select from it whatever was most important in the old mythology, and thus to compile a shorter and far more intelligible system.

But the sciences† here, as in every other country, have been subject to the greatest

* "Three pieces alone of this collection, though perhaps the best of it, have come down to us." *Northern Antiquities.*

† Von Troil.

revolutions, and, to use the words of Dr. Finneus (who, in his *Hist. Eccles. Islandiæ*, compares the state of literature in Iceland to the four stages of human life), their infancy extended to the year 1056, when the introduction of the christian religion produced the first dawn of light; their youth to 1100, when schools were first established, and the education and instruction of young men began to be more attended to than before; their manhood lasted till about the middle of the fourteenth century, when the sciences gradually decreased, and were almost wholly extinct, no work of any merit appearing. History now drooped her head, poetry had no relish, and all the other sciences were enveloped in darkness. The schools began to decay, and, in many places, they even had none at all. It was very uncommon for any one to understand Latin, and few priests could, with fluency, read their breviary and ritual.

The reformation produced in Iceland a new dawn of learning; and a few rays of that light which has blazed over Europe, from the discovery of printing, shed a gleam on this remote island. But it is to Bishop Areson, one of the most illiterate and bigotted of the Roman Catholic bishops, that the inhabitants are indebted for the introduction of the first printing-press. He, anxious to undermine the power of the king, and to hinder the progress of the reformation, but ignorant of the Latin language, which was made use of in letters of excommunication and other ordinances, commissioned a friend to procure him a person well versed in Latin, who might, at the same time, establish a printing-office. For this purpose, Jon Mattheson, a Swedish priest, was invited to Iceland, whither he conveyed a press, and fixed it in the district of Hunnevatn. At his death, his son removed it to Nupefell, where he printed some books at the time that Bishop Gudbrand Thorlacius* began to print at

* "Ille non modo suæ ætatis, sed et posteritatis ornamentum. Qui præterquàm quod inchoatum opus à prædecessore Olao sibi relictum ducente S. S. optimè ad eam, quam dedit Deus perfectionem, deduxit, (Dico labores et diligentiam in asserenda veritate Evangelica, et papisticis superstitionibus abrogandis) etiam in hac patria sua officinam Typographicam primus Islandorum aperuit. Cui ideirco patria inter libros

Holum. Bishop Thor Thorlaksen, in 1685, transported it to Skalholt, whence it was again restored to Holum by Bishop Biorn Thorlevsen. About the middle of the eighteenth century a new printing-office was established at Hrappsay, by Olaf Olssen; and hence, as well as from Holum, many valuable works have issued. Of late, however, the office at Holum has been suppressed, and the only one now in the island is situated at Leera, in the district of Borgafiord.

For an account of the present state of literature in the island of Iceland, I must refer to the fifth chapter of *Sir George Mackenzie's Travels in Iceland,* where Dr. Holland has amply treated on this subject. From it the limits of my Introduction will allow me to extract little more than the names of some of the most celebrated of the living authors.—Of such are Finnur Magnusen and Professor Thorkelin, who

complures in linguam vernaculam translatos, etiam sacrosancta Biblia, elegantissimis typis Islandica lingua in officina ipsius excusa, in æternum debebit." *Arngrim Jonas, Brevis Comment. de Isl.*

have made the early literature of Iceland the particular object of their studies; and Steingrim Jonas of Bessested; the Rector Hialmarson, who formerly conducted the school at Holum; and Arnes Helgeson, the driest of Vatnsfiord, who have distinguished themselves in classical knowledge. Assessor Benedict Grondal, a judge in the high court of justice, is mentioned as the most eminent among the poets, although his performances are almost wholly confined to odes, epitaphs, and other detached pieces, among which are many excellent translations from Theocritus, Anacreon, and Horace. Finnur Magnusen is likewise celebrated for the facility with which he composes in the Latin and Danish languages, and for the extreme accuracy of his Icelandic style*. Jonas Thorlaksen, the translator of Milton, has composed many original poems of great merit. Sigurdar Petersen of Reikevig, has written, among other things, a poem, in six cantos, called Stella, in which, under a fictitious form, the

* I have before alluded to his poem, inserted in the Appendix of the second volume of this Tour, and, at p. 39 of this volume, is noticed a translation of the Georgics of Iceland, into Danish verse.

manners and habits of the Icelanders are minutely described. Magnus Stephensen, the Etatsroed, is justly entitled to the first rank among the historical writers; and, in a list of his works, no less than twenty, on various subjects, are enumerated by Dr. Holland: many of them, however, are published for the use of a literary society, of which Mr. Stephensen is president. Numerous works on divinity have appeared since the time of the reformation; but, happily for Iceland, metaphysics do not appear to have occupied the attention of the Icelanders in a great degree. The sciences, strictly so called, Dr. Holland goes on to observe, engage but few votaries. In natural history * the *Enarrationes Historicæ de Natura et Constitutione Islandiæ* of Eggart Olafsen deserve notice; as do the *Travels in Iceland*, published by the same gentleman, in conjunction with his companion Paulsen; a work con-

* The authors of the *Voyage en Islande* make mention of a Latin work published one hundred and fifty years ago, entitled *Theatrum Viventium*, and they speak of *Jon Olafsen*, who flourished about the middle of the seventeenth century, and had made natural history his particular study. He travelled much in Europe and in the East Indies, and wrote an account of his life and travels.

taining a vast store of information, but miserably deficient in arrangement. Olaf Olafsen printed, in 1780, his *Œconomical Travels through Iceland,* containing much valuable matter. Jon Soemundsen has written on the volcanic eruptions that have happened in the neighborhood of the lake Myvatn; and Bishop Finnsen on Hecla; and Mr. Stephensen's *Account of the Eruption of Skaptefield Jökul* will be found translated into English, in the latter end of this journal.

Mathematics and astronomy are but little cultivated, though the elder Mr. Stephensen and Stephen Biornsen have written on these subjects.

In the fine arts no progress whatever has been made; but, as a proof that this deficiency is rather to be ascribed to the situation of the people, than to a want of original genius, Dr. Holland remarks, that Thorvaldsen, the son of an Icelander, dwelling on the classic ground of Rome, is second only to Canova among the statuaries of Europe.

The remains of antiquity in Iceland are few and of small importance, since the country has been plundered of all its old manuscripts. Of ancient edifices scarcely any traces remain; for the mode of building practised in the island with pieces of rock without cement is of itself naturally unfavorable to the duration of the walls, and has also greatly facilitated the attempts of the natives to take them in pieces as often as they wanted the materials to erect others. The mere foundations of large structures are alone now and then to be traced, one of which that served as a pagan temple is distinguishable by the Blodstein, or stone for sacrifice, which is of an oval form, a little pointed at the top.

Equally insignificant are the ancient inscriptions that have been found in the island: the most remarkable among which is that at Borg, in Myrar, the epitaph of one Kartan, a man of regal extraction, who fell by the hands of an assassin. It is engraved in Runic characters upon a kind of rock resembling basalt.

Some fragments are still preserved of the armour of former days, such as a halbert, long kept in the cathedral of Skalholt; and a few swords, with a lance and helmet, which are to be seen at Hlidarende; but they are said to possess nothing remarkable in their form. Sepulchral monuments, consisting of heaps of stones, resembling the cairns of Wales and Scotland, are scattered in small quantities over the island.

The principal exports of Iceland are dried fish, mutton, lamb and beef, butter, tallow, train-oil, coarse woollen cloth, stockings, gloves, raw wool, sheep-skins, lamb-skins, fox-skins, eider-down, and feathers, to which in former times was added sulphur. They import timber, fishing-tackle, various implements of iron, tobacco, bread, spirituous liquors, wine, salt, linen, with other necessaries of life for the people in general, and a very few superfluities for the richer class of inhabitants. At its earliest period Iceland appears to have been the rendezvous for all the disaffected and discontented among the Norwegians and Danes, and was little more than a nest of pirates; but after the island had submitted to the Kings of Norway, and a security was afforded to commerce, the vast quantities of wool, tallow, oil, and other products that were exported, brought back so large a return of the precious metals, that it was reckoned a desirable situation for adventurers to make their fortunes in. Many concurrent circumstances afterwards occasioned the decay of this trade, but nothing so much as the king's usurping the whole commerce of the island, and affixing certain prices to all the produce; so that no man dared to sell any thing, except to the royal factors, nor to them at a price above what was stated in a printed list that was circulated all over the island. A monopoly of this nature at first produced great revenues to the royal treasury, but the people soon became impoverished by it, and, following the natural course of things, the factors began to oppress the natives and to cheat their master, so that at last the profits were not equal to the expence of such a commerce. The Danish government therefore issued proclamations *, declaring the trade of Iceland to be free. But, if the island had suffered formerly by the factors, it suffered

* The nature of the Proclamations relative to the freedom of Trade will be seen at Appendix F. of this work, where some of them are translated.

much more by the measures that were now adopted; for this nominal freedom consisted in the king's privileges being sold to a body of merchants, who enjoyed, under certain stipulations, the exclusive right to trade with the island. The natives were under the same restrictions as before, nor could any ships, but those of this company, come into the Icelandic ports to traffic. The principal purchaser from the king did an essential injury to the inhabitants, by suffering the manufactory of cloth to go into decay, whereby numbers were exposed to poverty and want. He was, by so doing, able to export the raw wool to a greater profit, and also to have a farther advantage by importing cloth and other manufactured goods.

In nothing do the Icelanders excel so much as in the curing of the cod-fish *,

* The following particulars relative to the curing of the cod-fish, extracted from the *Voyage en Islande,* may not be unacceptable to my readers." —1º —Ils enlèvent à la morue l'épine dorsale (qu'ils appellent *Blod-Dalken*) jusqu'à la troisième vertèbre, au-dessus du nombril: cette opération leur est même ordonnée. Elle fait que le poisson se sèche plus promptement, et

which is of the best kind; so that, if the fisheries were properly conducted, they might prove a source of inexhaustable wealth to the

que l'air pénètre mieux dans les parties où la chair, est épaisse. On a soin en même temps de bien faire saigner le poisson, afin qu'il ne se noircisse pas, et ne prenne pas un mauvais goût, ce qui en empêcherait la vente. Aussi les pêcheurs expérimentés éventrent-ils les morues qu'ils pêchent, dès l'instant où ils les ont tirées dans la barque; et ils les percent tout prês de la tête jusqu' au cœur, ce qui fait tout-à-coup écouler la totalité du sang qui est encore fluide. Elle acquierent par-là une blancheur sans égale et la chair en devient très-belle et très-appétissante.—2º —On fait sécher les têtes, parce qu'on ne les mange que rarement dans leur fraîcheur. —3º —On se met ensuite à preparer la vessie (qu'ils nomment *Sundmaven,* et les commerçans étrangers *Sunnemave*) qui consiste en une peau coriace, semblable à du cuir. Elle a une ligne d'épaisseur; elle est parfaitement blanche et pleine de ligamens ronds et creux, qui la tiennent attachée aux côtes. Elle est située sous l'épine dorsale, dans le ventricule du milieu, et est communément plein d'air. Comme on met beaucoup de dextérité et de promptitude à tirer le poisson de l'eau, dès qu'il a mordu à l'hameçon, cette vessie se gonfle tellement par les secousses, que s'il vient à se détacher, il flotte longtemps sur l'eau, et demeure un certain temps dans cet état, avant de pouvoir y redescendre. Si on éventre le poisson aussitôt qu'il est pris, et qu'on perce un trou

island; for fish from that country always sells at a much higher rate than what comes from either Newfoundland or Norway.

dans la vessie, l'air en part avec impétuosité et une éspéce de bruissement; si l'on amène au contraire une morue ave douceur dans la barque, c'est-à-dire sans secousse, la vessie ne s'enfle nullement. Dans une morue maigre, qui aura resté quelque temps sous l'eau, sur un terrain ou fond argilleux, on trouvera cette vessie pleine d'une matière visqueuse et jaunâtre, mais passablement liquide. Elle forme un mets agrê-able, sain, léger, et nourrisant. On s'en sert ici et chez l'étranger au licu de colle. Plusieurs la confondent avec celle qu'on fait de la vessie de l' Icthyocolle, inconnu dans ces parages. Les gens gagés ou domestiques, que l'on envoye à la pêche, sont chargés, en vertu d'une ordonnance de police, de la préparation du corps de la morue, de celle de la tête, de la vessie, du morceau qui forme le chignon et de l'extraction de l'huile: ils sont tenus d'en rendre compte à leurs maîtres. La morue sèchée en plein air est excellente et d'un gôut si agréable, que plusieurs la préfèrent à celle qu'on séche sur les rochers *. Pour la sécher de cette manière, on s'y prend de bonne heure au printemps. Lorsque le vent est nord, elle acquiert communément beaucoup de blancheur au dehors et ses filamens deviennent comme frisés et rudes; dans

* Ils appellent le poisson qu'ils font sécher sur les rochers *Klipfisch*, au lieu qu'ils donnent le nom de *Fredfisch* à celui qu'ils séchent en plein air, c'est-à-dire, suspendu à des cordes.

With regard to the amusements of the Icelanders they are not of a kind calculated to dispel the gloomy habit which continually hangs about them; and, indeed, they are now almost entirely confined to the reading or repeating one to another their ancient *sagas:* these are the delight of the youth as well as of the aged; but while the more authentic manucript histories of former times are the means of enabling them to retain and speak their language in its almost original purity, the mere traditionary ones are replete with

l'intérieur, la chair devient au contraire rouge et tendre. Lorsqu'on frappe ce poisson pour l'applatir, il se trouve un certain déchêt, par rapport aux filamens externes qui tombent en poudre; mais on n'y fait pas attention, dans des parages où il est aussi abondant. On voit néanmoins des personnes économes qui en tirent profit; elles ramassent cette poudre ou farine de poisson, et s'en font un mets délicat et d'une facile digestion. Cette poudre perd ce goût fort et âcre, que le poisson acquiert lorsqu'il est sec, et qui le rend peu propre à être vendu aux commerçans, quoiqu'il perde par la désiccation une bonne partie de son suc. Cependant on n'a pas encore décidé, si un poisson séché à la gelée n'est pas plus sain que celui qui l'a été dans tout autre temps, puisqu'elle lui enlève toutes les parties visqueuses et aqueuses, et ne lui laisse que les parties grasses et salines."

absurd stories, that keep alive a love of the wonderful, and impress with superstitious notions the minds of almost all the lower class of people. In former times wrestling and various feats of strength used to occupy their attention; chess was much practised; and cards, music, and dancing diversified their leisure hours: but all these are now scarcely heard of. Their attachment to their native land is very strong, and might be accounted truly wonderful, since the country seems entirely destitute of every thing which can add to the comforts of life, and nearly so of the means of procuring a necessary subsistence, were it not that, "Providence," as Von Troil well remarks, "has wisely in-"stilled into the human heart, the love of "that soil whereon a man is born; and, "probably with a view that those places "which are not favored by nature with her "choicest blessings, may not be left without "inhabitants, it may be affirmed with some "degree of certainty that the love of one's "native place increases in an inverse ratio "with its having received favors from na-"ture." This is, indeed, most justly applicable to the patient and contented Icelander;

who, happy in the lot that Providence has assigned to him, is scarcely ever known to leave his cold and barren mountains for all that plenty and comfort can offer him in milder regions *.

* The first settlers, however, who were famed for their maritime enterprizes, had more of a roving disposition. Torwald was induced to attempt the discovery of a coast to the north of Iceland, before seen by one Eric Rufus. In the year 928, he made good a landing, and, having surveyed it, he gave it the name of Greenland. After living there some years he returned to Iceland, and prevailed on several persons to go and settle in this new country. Two towns, Garde and Albe, were founded; a monastery was established and dedicated to St. Thomas, and all the inhabitants acknowledged the Kings of Norway for their sovereigns. This colony subsisted till the year 1348, when the dreadful pestilence, called the *black death*, committed its ravages, and from that time these settlements seem to have been wholly forgotten or neglected, though Egede, in his *History of Greenland*, offers proofs that the whole colony is not wholly extinct, and even proposes means of getting to it. It was in one of these voyages to Greenland that an Icelander, named Biarn, driven to the southward in the year 1001 by tempestuous weather, discovered land, flat and covered with wood, which it has since been supposed must have been either Labrador or Newfoundland; this was again visited by some of the inhabitants of Greenland, who gave it the name of

The employments of each individual Icelander are necessarily various, since artists, mechanics, and people of different professions are almost unknown among them. *In the winter the care of the cattle is of the highest importance; the stoutest and most healthy of the men are then occupied in the preservation of those to which shelter and dry food cannot be afforded at this inclement season, and it is necessary to remove the snow as much as possible from the grass, that the beasts may be able to procure a subsistence, however scanty. Other men are employed in picking the coarse wool from the fine, and manufacturing it into ropes, bridles, stirrup-straps, and cushions, which are often used instead of saddles. They

Vinland, and established a small colony, whither many persons both Greenlanders and Icelanders resorted. But as a more detailed account of the discovery and settlements in these two places, although connected with Icelandic history, would carry me beyond the intended limits of this Introduction, I will beg leave to refer my readers to the first volume of *Percy's Northern Antiquities*, for much more interesting information on this subject.

* *Voyage en Islande.*

also prepare skins for their fishing-dresses, and tan others to make into saddles, as well as thongs to fasten burthens upon their horses, and they forge iron into scythes, horse-shoes, and different kinds of tools. The women find abundant occupation in washing the wool, and in picking, carding, and spinning it; as well as in knitting gloves and stockings, and in weaving or dying flannel and stuffs for their various dresses, all which they make themselves; as they do their shoes of untanned skin. The fulling of the cloth falls to the lot of the men.

As early as the month of February or March, the fishing-season calls the men or at least the greater number of them to the coast: others only resort thither in the summer, when the fishing is nearly completed, and take with them their butter and wadmal to exchange for the fish, with which they return loaded. At that time of the year, also, the Danes are accustomed to arrive in the different ports, and an opportunity is thus afforded to the natives of carrying on a little trade with them. To the fishery succeeds the season for drying and securing the hay, and another migration takes place of the poorer inhabitants from various parts to assist the farmers. The salmon-fishery and the cutting and preserving of turf for winter fuel are at the same time attended to.

In the autumnal months the necessary repairs are done to the dwellings, the grass-land is manured, and the sheep are killed and cured either for winter store or for exportation.

The more industrious exercise their ingenuity during their leisure hours in the manufactory of various articles in brass, silver, and wood, such as girdles, buttons, clasps, ornaments for their saddles and dresses, snuff-boxes, &c.; in all of which they display an extraordinary neatness and elegance of workmanship. Some of them, too, are excellent boat-builders. The women embroider their garments with figures of flowers and animals of various forms and colors.

The principal articles of food among the Icelanders are fish and butter; the former mostly eaten in a dry state and uncooked;

the latter made without salt, with all the whey and superfluous moisture pressed out, in which state it will keep for fifteen or twenty years, acquiring in the interim a degree of rancidity which is not unpleasant to an Icelandic palate. During the time of the prevalency of the Popish religion *, a large building was appropriated, at each of the episcopal sees, for the purpose of laying by a store of this butter, which was packed down in chests, each thirty or forty feet long, by four or five feet deep, and was thence distributed among the most necessitous of the natives in seasons of famine or scarcity. Milk is converted into *Syra*, or sour whey, which is preserved in casks, till it has undergone the process of fermentation before it is used as a beverage. The same mixed with water is called *Blanda*. *Striuger* is whey boiled to the consistency of curd; and *Skiur* the same from which the liquid has been expressed. The flesh of either sheep or bullocks and rye-bread are only brought to the table of the superior class of people. Birds of various kinds,

* *Voyage en Islande.*

especially water-fowl and the larger inhabitants of the deep, are of course but occasionally procured and cannot be taken into account, while speaking of the general mode of subsistence of the Icelanders, any more than the native vegetable productions which are occasionally prepared for food; such as the *Angelica Archangelica*, *Cochleariæ*, *Rumices*, and *Dryas octopetala*, with *Lichens* and *Fuci* of two or three kinds. The *Lichen islandicus* alone is sometimes eaten in considerable quantity; but more is gathered for exportation.

Of the amount of the population of Iceland in early times I am ignorant, except as far as some sort of estimate may be made from what is mentioned by Arngrim Jonas*, that four hundred people paid tribute in the year 1090; but in this number neither women, children, nor poor were included. In the fourteenth century a dreadful malady† called the *sorte dod*, or black death, is reported to have swept away almost every in-

* Arngrim Jonæ Brev. Comment. de Islandiâ.

† Horrebow.

habitant from off the island; so that, comprehensive as are the annals of Iceland, this circumstance is omitted in them, and it is thence inferred that no person of ability survived to record it. The years 1697, 1698, and 1699 were remarkable for the mortality caused by famine, and the year 1707 for the destruction of twenty thousand inhabitants by the small-pox; yet in 1753 Horrebow estimates the population at eighty thousand, and Von Troil in 1772 at sixty thousand; but, in consequence of the tremendous eruption of Skaptar-Jökul in 1783 and other unfortunate events, the number is now reduced to forty-eight thousand. Independently of the destructive effects of volcanoes, disease, and famine, which so often ravage the island, the quantity of those who die in their infancy for want of proper nourishment is extreme. It is remarked* that Barderstrand Syssel in the year 1749 contained three thousand inhabitants, but that in the short space of thirteen years (in 1762) this amount was diminished to two thousand one hundred and seventy-five. From the poverty of this district the want of necessary nutriment for

* *Voyage en Islande.*

young children is increased, and two-thirds of the number born are supposed to perish in the cradle. It seldom happens that out of twelve or fifteen children, which the women sometimes produce, one-half of them live, and more commonly only two or three are brought up to manhood, though most of those survive that are preserved through their first or second year. What makes this period so peculiarly fatal, is the custom that prevails among the women of not suckling their infants at all, or at most only for a few days, after which they feed them with cow's milk, which is taken through a quill with a piece of rag fastened to one end for the sake of softness to the mouth *.

The Icelanders in general do not attain to an advanced period of life, though many live to the age of seventy and enjoy a good state

* A similar method of feeding infants is mentioned by Linnæus, in his *Lachesis Lapponica.* When he was in Lycksele Lapland, he says, "I remarked that all the women hereabouts feed their infants by means of a horn, nor do they take the trouble of boiling the milk which they thus administer, so that, no wonder the children have worms. I could not help being astonished that these peasants did not suckle their children". *v.* I. *p.* 178.

of health; but this is among the higher class of people. The nutriment of the poor and their manner of living is unfavorable to longevity, independently of the dreadful cutaneous diseases to which they are subject. Scurvy, leprosy, and elephantiasis are no where, perhaps, more prevalent; and they are likewise, according to Von Troil, peculiarly afflicted with St. Anthony's fire, the jaundice, pleurisy, and lowness of spirits.

The climate of Iceland is not so settled as that of equal latitudes upon continents. In the winter the inhabitants are exposed to frequent and sudden thaws, and in the middle of summer almost as much so to snow, frost, and cold, so severe as effectually to prevent all cultivation. The year 1809 was particularly unfavorable: I recollect that in the early part of that summer Fahrenheit's thermometer varied in the course of the day from about 41° to 45°, seldom rising to 50°, and only once to 60°. Mr. Savigniac, however, assured me, that at Reikevig one day the thermometer, exposed to the sun, rose to 100°. In the beginning of August there were severe frosts, and much snow fell in the vallies and plains, even in the most temperate parts of the island. In common seasons * the changes that take place in the atmosphere in the course of the twenty-four hours are very extraordinary; since it often happens that after a night of hard frost the thermometer will in the day rise to 70°. During the winter of the year 1348, the annals of the country relate that the sea was frozen all round the coasts, and that a person might ride on horseback upon the ice from one cape to another across all the gulphs and bays in the island. In February, 1755, the thermometer in the southern quarter of the country, fell to 7°. In 1754, on January 13th, it was at 9°; on February 13th, 8°; on the 14th of March 11°; on December 6th, 11½°; and on the 12th of the following February, 12°; even in the month of May, in the same year, the frosts were so severe that in one night's time water in the neighborhood of the sea was frozen an inch and half in thickness. Ice-islands in the years 1615, 1639, 1683, and 1695 came round to the south coast, which is by no means an usual circumstance.

* *Voyage en Islande.*

The northern part of the island is, as may be concluded, exposed to much more severe weather than the southern*. Vegetation is scanty, and the herbage difficult to be dried for hay. The quantity of floating ice driven by the westerly and north-westerly winds from the coast of Greenland is prodigious, and not only fills all the bays, but covers the sea to that extent from the shore that the eye cannot trace its boundary from the highest summit of the mountains. These masses of ice, known by the name of ice-islands, are so large that a body of sixty or eighty fathoms in thickness is sunk below the level of the water, and a height of many toises rises above it. Their motion is rapid, and they are often driven together by the sea with so tremendous a crash that the report is heard at an immense distance, and with such force, that, according to Povelsen and Olafsen, the pieces of float-wood that they bring with them, have been known to take fire, in consequence of the friction. It is a singular fact, that, so long as these ice-islands continue floating

* *Voyage en Islande.*

about in the ocean, the weather is fickle and stormy, and the current and ebb and flow of the tide are all in disorder and confusion: but, as soon as they become stationary in the gulphs and inlets, and the waters have carried away the smaller detached pieces, nature returns to its accustomed state of order and regularity; the weather growing calm in the country, and the air thick and loaded with fogs, though at the same time accompanied by a moist and penetrating cold. Among the inconveniences arising from the arrival of this ice, besides the excessive cold which destroys vegetation and cattle, is to be reckoned the opportunity it affords for the white bears of Greenland to visit the country, which they occasionally do in alarming numbers, and render it necessary for the natives to assemble in parties for the purpose of destroying them, lest so unwelcome a visitor should fix himself permanently among them.

In mentioning the general face of the country, I cannot do better than copy the exclamation of Von Troil on his arrival. "Imagine to yourself an island, which from one end to the other presents to your view only barren mountains, whose summits are covered with eternal snow, and between them fields divided by vitrified cliffs, whose high and sharp points seem to vie with each other to deprive you of the sight of a little grass which scantily springs up among them. These same dreary rocks likewise conceal the few scattered habitations of the natives, and no where does a single tree appear which might afford shelter to friendship and innocence. The prospect before us, though not pleasing, was uncommon and surprising. Whatever we saw bore the marks of devastation, and our eyes, accustomed to behold the pleasing coasts of England, now saw nothing but the vestiges of the operation of a fire, heaven knows how ancient!" Of the mountains of Iceland, some are composed of loose fragments of rock to their very summit, while others apparently retain their primæval form and nature, lying in horizontal strata. The height of a very few has been accurately ascertained; and these, though said to measure nearly seven thousand feet of elevation, are by no means the loftiest in the island. Geitland and Blaa-fel Jökul tower over the rest in the southern quarter, where Hecla, also, is situated, more remarkable for the frequency of its eruptions than for its height, which is only about five thousand feet. The western quarter of the island contains, among other vast mountains, Snoefel Jökul*, well known to all navigators along that coast, more by its vicinity to the sea, than its great elevation; and Boula, conspicuous for its singularly conical form. Lange and Hofs-Jökul are the loftiest in the northern division of the country; and in the eastern Klofa, Skaptar, and Torf Jökul, the latter esteemed the most stupendous in the whole island.

Rivers and fresh-water lakes abound; the latter of very considerable extent and well

* Snoeful Jökul, which I have in the course of my Journal, stated, upon the authority of Eggert Olafsen, to be seven thousand feet in elevation, has been ascertained by Sir George Mackenzie to be only four thousand five hundred and fifty eight feet high. His observation is also confirmed by the calculations of the two Danish officers who are employed in surveying the coasts.

supplied with fish; the former, though of sufficient width in many instances to admit of navigation, are too much obstructed by rocks and shallows to be employed to this important object. The bays and harbors are both numerous and safe, though their entrances are but little known, except by those who are frequently in the habit of visiting the coasts.

The annals of the island describe the country, than which nothing can possibly be now more bare, as having been once covered with impervious forests; and the quantity of bog-wood and *surturbrand* which is continually dug up affords the most decisive proof in favour of the truth of such assertion. Even now, too, the name remains, though the reality has long ceased to do so, and places are called forests that produce only a few miserable and stunted birches. All attempts of recent times to cultivate even the most hardy trees have proved ineffectual, so that for his necessary supply of wood the Icelander is obliged wholly to depend upon importation from Norway, excepting only what he gets from the northern and eastern coasts of his own island, where much timber is frequently cast by the waves of the sea, conveyed, as it is supposed, by the winds and currents from North America.

The natural history of the island, its volcanoes, its sulphur-springs, and its boiling fountains, are spoken of so much at large in the Journal and Appendix that it is needless in this place to mention them. Those who may be desirous of more information on any of the points here glanced at, I beg to refer to the able works of Von Troil and Povelsen and Olafsen; for these pages, to use the words of the most popular poet of our days, "are but a tale of *Iceland's Isle,* and not a history."

Halesworth, December 9, 1812.

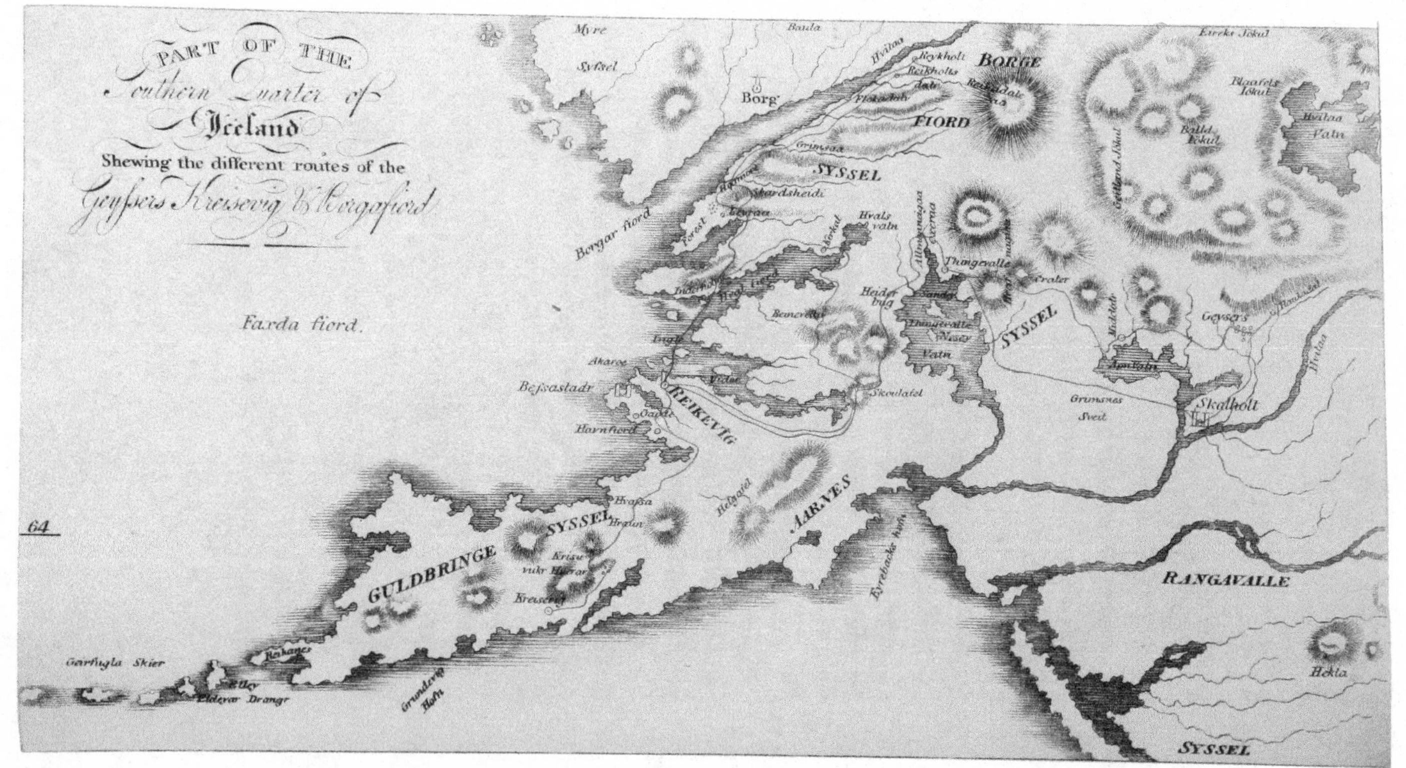
PART OF THE
Southern Quarter of
Iceland
Shewing the different routes of the
Geysers Kreisevig & Borgafiord
64
Faxda fiord.
Myre
Syssel
Baula
Borg
Hvitaa
Reykholt
Reikholts dale
BORGE
FIORD
SYSSEL
Grimsaa
Skardsheidi
Borgar fiord
Forest
Hvals vatn
Thingevalle
Crater
Heider bug
Thingevalle Vatn
SYSSEL
Geysers
Eireks Jökul
Blaafels Jökul
Balld Jökul
Hvitaa Vatn
Bessastadr
Akaroe
REIKEVIG
Skoudafel
Grimsnes Sveit
Skalholt
Hvitaa
Havnfiord
Hvassa
Hraun
Helgafel
AARNES
Eyrebacks havn
RANGAVALLE
Hekla
SYSSEL
GULDBRINGE
SYSSEL
Kreisevig
Reikanes
Gierfugla Skier
Eldey
Eldeyar Drangr
Grundevig Havn

RECOLLECTIONS

OF

ICELAND.

1809. Friday, June 2. EARLY this morning, the Margaret and Anne, Captain Liston, bound for Reikevig in Iceland, being ready for sea, and my luggage having been previously sent on board, Mr. Phelps, Mr. Jorgensen, and myself embarked from Gravesend. From the excellent accommodation which the vessel afforded, and the pleasant society of the two companions of my voyage, I flattered myself, and not in vain, with as agreeable an excursion as the nature of the circumstances would allow. Friday, however, being considered by all sailors as an unlucky day to commence a voyage, our people were so tardy in their preparations to get under way, that, before noon, a violent hurricane, which came on and continued all day, obliged us to keep our station; at least, as much so as the storm would permit; for we dragged our anchors a considerable distance. The howling of the wind among the rigging, joined to the sight of a number of large vessels that were driven on shore, and of boats in distress in every direction upon the river, did not strike us with very pleasing sensations, although we were riding in perfect safety: to add to the scene, a house close to the shore was discovered to be in flames. Towards evening the storm abated, and early the next morning, with a charming breeze, we sailed down the river; and, while looking with delight on the green and fertile shores, we thought of the far different appearance of those more striking scenes of fire and devastation, which Von Troil, in his Letters on Iceland, had taught us to expect in that country. In the after part of the day the wind increased, and, towards night, blew so violently, that our captain thought it most prudent to come to an anchor in Hollesley Bay, and, in the morning, to fire a signal for a pilot. When the violence of the storm had

Saturday, June 3.

Sunday, June 4. abated the next morning, a boat came off to inform us, there was no pilot to be had; therefore, with a more frequent use of the lead, the captain determined to pass the sands off the coasts of Norfolk and Suffolk without one. Having cleared these, we steered more to the eastward, and soon lost sight of land. When we were about seventy miles from the shore, a Willow-wren, *Motacilla Trochilus*, was observed flying about the rigging of our vessel, and, soon after, a female Black-cap, *Motacilla atricapilla*; and, when we were still further out at sea, *Hirundo domestica* and *H. Apus* were skimming about us, and every now and then resting upon our ropes. These birds were probably driven from our own coasts by the late strong westerly gales, as it is too late for the regular migration of these, our summer visitants. On the third morning of our losing sight of land, *Hirundo urbica* settled upon our rigging, and seemed much fatigued. We had, from the time of our leaving Hollesley Bay, so fine and so favorable wind, that on the Wednesday Wednesday, June 7. morning I was called from my cot, and asked if I knew the coast which was in sight. I immediately recognised Wick Castle, in Caithness, standing upon a rocky peninsula, and we soon descried Wick itself. From Sleigo, an adjoining village, we took on board two pilots, and, with great rapidity, passed the three extraordinary conical and insulated rocks, called the Stacks of Duncansby. Here, we calculated that we had run seven hundred miles, and six hundred and twenty-two of these in three days. The Orkney Islands were, in a great measure, obscured from our view, as we dashed through the Pentland Frith, by a thick fog, in which most of them were enveloped. We could, however, distinguish Stroma, South Ronaldsha, and Hoy, and, in the latter island, the hospitable seat of Colonel Moodie, at Melsetter. Opposite to it, on the Caithness coast, whilst viewing the venerable residence of the Earls of Caithness, I recollected the hearty welcome and kind assistance that Mr. Borrer and myself received there, from the present hospitable proprietors, but nine months before, when we were rambling over these northern parts of Scotland. Near to Mey Castle was the Church of Caninsby, and, on the opposite side, the steep cliffs of

Dunnet Head. When we had got out of the frith, a clearer atmosphere gave us a view of the Paps of Caithness, in the south-east: in the south, was the fine conical hill, called Ben-y-Græme; and, in the south-west, the great mountains of Ben Hope and Ben Luyal, in Sutherland. No sooner had we passed the frith, and got to the westward of the Orkney Islands, than we were becalmed, and continued so for two days; Hoyhead and the Old Man of Hoy, a singular rock near the shore, being most of the time in sight. On the

Thursday, June 8.

Thursday, a *Tringa*, which appeared to me a new species, flew on board quite exhausted, and was taken. From this time calms or bad winds, and frequently, heavy squalls, attended us, so that we made but slow progress. About the hour of mid-

Wednesday, June 14.

night, on the 14th, we descried land in the horizon, or rather snow, for, as we approached it, we could discover nothing but mountains of prodigious magnitude, covered on every side with snow, and most distinctly seen, from being backed by a dark cloud, though at the distance, as we computed, of fifty miles. On the highest ridge of these mountains were some huge angular and projecting precipices, which cast a deep shadow on the white snow, when the early rays of the sun were striking upon them, breaking the uniformity of such an extended outline. This range of mountains we afterwards discovered to be Klofa Jökul (Jökul, pronounced Yuckul, meaning a range of snow mountains), in the south-eastern part of Iceland, and Mr. Phelps and I gazed upon it with astonishment and delight, till a late hour in the morning. Such a scene was quite novel to us, and the circumstance of our contemplating it all night long did not at all diminish its effect. To the north-east of this, we saw a long stretch of nearly level land, of, comparatively, no great elevation, but every where covered with snow, and only here and there interrupted by a rugged mountain, whose sides were of a very rude figure. The following night, we passed within sight of a flat extent of land, which appeared to be about twenty miles from us, and on which, by the help of our glasses, we could plainly discern a number of buildings; but we could never learn what place it was. I do not recollect ever after, except at Reikevig, seeing so many

Thursday, June 15.

Friday, June 16. houses together in Iceland. About two o'clock the next morning, we discovered Westman's Isles, or Vestmanna-eyer, as the natives call them. These islands are so named, from the circumstance of some Irish fugitives, who had killed their master, having escaped to them in A.D. 875; for the Irish and Scotch were both called by the ancient Norwegians *Vestmen*. According to Povelsen and Olafsen, some places in the principal, and the only inhabited, island, are still known by the names of these Irish runaways. The whole groupe appears perfectly barren, and they rise to a vast height, and of the strangest shapes, perpendicularly from the sea. We had a magnificent view, as we passed close by them with a light breeze, which, however, was scarcely sufficient, in our captain's opinion, to take us out of the force of the currents, which run here with great velocity. As we proceeded, the different sides which came to our view presented different shapes and appearances; in some, these sides hung over the deep, as if they would fall every instant; others had a perforation at their bottoms, through which a boat under sail might pass: all of them were of a dark brown color, but whitened in places by the dung of the immense quantity of birds which constantly frequent them. In the afternoon, we saw other Jökuls, which were covered with snow, and extended in uninterrupted lines almost as far as our sight could reach. Hence, we bore a little more to the southward, in order to double a dangerous chain of rocks running out from the south-east corner of Iceland, and called the Fugle Skiers. We soon lost sight of our snow mountains, and, instead of feasting our eyes with these wonders of the northern regions, had to encounter three days of almost incessant squalls, sleet, rain, and a most boisterous sea. When, at length, we supposed we had sailed far beyond the outermost rock (for we gave it a birth of twenty-five miles), we steered to the northward, and reckoned the next morning upon entering the great bay of Faxa-fiord. We were all thrown into confusion, however, by Mr. Jorgensen's accidentally looking out a-head, and discovering, within a few minutes sail, some breakers dashing over a sunken rock directly in our course. He immediately gave orders for putting the vessel about, and flew himself, with the greatest alertness, from one part of the deck to another, to assist, by his own exertions, where fear or hurry prevented the common sailors from doing their duty. Although it blew a gale of wind, so that, in getting about, our decks were completely washed by the seas, yet, it was done so rapidly, that no one, except Mr. Jorgensen, knew the extent of the danger, till we had escaped from it. Unfortunately, almost at the same time the wind shifted, and we were obliged to beat about to the southward for two days, before we could get round this dangerous reef, which was not laid down in any of our charts. Tuesday, June 20. At length, with more favorable weather, on the 20th we entered Faxa-fiord, and steered pretty nearly due east, to get into Reikevig Bay. On our right was a long flat extent of land, which is called Guldbringue Syssel, or District: from it, rose several insulated mountains, and one of a remarkably conical figure, but none of any great height. Wednesday, June 21. Early on the following morning, as we continued our course, other larger mountains came in view, on the mist's clearing away; and, after an interval of several hours from the time of our firing the signal, we saw a boat, with some pilots on board, approach us. We were delighted at seeing some new faces, in spite of their nastiness and stench; and their grotesque appearance afforded us much amusement. I cannot say that I observed any thing strikingly peculiar in their features: their faces were mostly broad, and, as to color, none of the fairest. Their stature was in general small, but one or two of them were rather tall, and, I think, not much less than six feet high. Some had pretty long beards, while others had as much only, as would remain after the operation of shaving had been performed with a blunt knife, or a pair of scissars: as to their hair, it was altogether in a state of nature, untouched by a comb, and hung over their backs and shoulders; matted together, and visibly swarming with those little vermin, and their eggs*, which are

* Much, and universally as the common people of Iceland are infested with these troublesome creatures, and greatly as they are sometimes distressed for food, I never saw or heard of their applying them to that use, which Kracheninnikow observes is common among the Kamtchadals, of whom he says, " Ces peuples sont remplis d'une si grande quantité de vermine, qu' en soulevant leurs tresses, ils ramassent la vermine avec la main, la mettent en un tas, et la mangent." Vol. i. p. 21.

the constant attendants of that part of the human body, when cleanliness is neglected. Their dress was simple enough, and warm; it consisted of a woollen shirt, a short waistcoat, and a jacket of coarse blue cloth or wadmal, and still coarser trowsers of the same materials, but undyed: the buttons were mostly of horn, and were, probably, from Denmark. Their stockings were of coarse worsted, and their shoes made of seal or sheep skin. Their gloves, too, were of the same materials as the stockings, that is to say, knitted worsted, made without divisions for the fingers, but having two appendages on each of them for the thumb: by this contrivance, when a boatman, in rowing, feels his hands galled, from the inside of his glove being wet and dirty, he turns the glove on the same hand, and has a dry and clean side against the palm. An Iceland hat is well contrived to keep the rain from the neck and shoulders; for it is furnished with an immense brim, which hangs down behind, in a manner not much unlike that which our London porters to the coal vessels make use of, but is equally large before. This, and the buttons, appeared to be the only articles of their dress which were of foreign manufacture. In the common conversation, which they held, in the Danish language, with Mr. Jorgensen, they seemed to be much animated, and had a great deal of action with their hands and heads; but as often as any thing was said or offered to them which gave them pleasure, they made it apparent by scratching and rubbing themselves violently, and writhing their body so as to cause it to chafe against their clothes; thereby indicating that they were sorely afflicted with a complaint, said, with what truth I shall not pretend to determine, to be very troublesome in the northern parts of our own island. These poor creatures swallowed the provisions that we gave them, with a most voracious appetite, and, by means of their excellent sets of teeth, our hardest biscuits were soon reduced to a digestible state. With our snuff and tobacco* they were highly pleased, and

* This passion for snuff and tobacco is prevalent among all the northern nations. I had frequent opportunities of observing it, during my tours in the Highlands of Scotland; and Linnæus has some curious remarks on the subject, in his *Flora Lapponica,* where he says, p. 310, "Ceterùm apud innocentissimos Lappos innotuit pessimus mos pulvere Nicotianæ nares

even boys of fourteen or fifteen years of age held out their hands for a piece of tobacco, whilst I was dividing some amongst the men. They invited us in their turns to partake of their snuff, but when they presented their boxes, we were at a loss

saturandi, ut nec vir nec femina nec puer sit, cui non in bursâ adsit pyxis pro pulvere olfactorio tabaci, pro tabaco conscisso ad suffumigium, pro comminuto ad morsulas. Sed notes velim condimenta; non enim simplex pulvis Nicotianæ sternutatorius sufficit naso ditiorum, sed pulvere Castorei saturatus erit, quo vehementius, gratius, salubrius spiret, licet nullam in Lapponiâ hystericam viderim; hinc in loco natali venditur communitèr integer folliculus Castorei tribus florenis, vel, quod idem, unico rhenone." This custom, however, is not confined solely to northern countries; for, in the town of Leetakoo, in Africa, in latitude 26° 30′′ south, (according to the account written by some travellers who penetrated into that country, and published by Mr. Barrow,) the practice of snuff-taking is said to be peculiarly agreeable to the natives. "This article is composed of a variety of stimulant plants, dried and rubbed into dust, which is usually mixed with wood ashes; of this mixture they take a quantity in the palm of their hand, and draw it into their nostrils through a quill, or reed, till the tears trickle down their cheeks. Children, even, of four or five years of age, may be observed taking snuff in this manner." *Voyage to Cochinchina*, p. 395.

how to get at a pinch; for their boxes* are shaped generally like a small flask, with an extremely narrow neck and mouth, which is stopped by a plug or peg of wood, fastened by its upper end to the neck of the box by means of a piece of string. The sides are carved with ornaments of various kinds, and inlaid very neatly with brass or silver: at the bottom, by means of a larger hole, which is closed by a screw, the snuff is admitted into the box, and our pilots soon shewed us their method of getting it out for use, which was, by holding their heads back, and inserting into one of their nostrils the mouth of the box; when, by two or three gentle shakes, a sufficient quantity is admitted into the nose, to produce the desired effect. Nothing more was then required, but to wipe away the superfluous particles from the nose, by drawing the back of the hand across it. However, this is not the only, although the ge-

* Their shape might, perhaps, be more aptly compared to a pair of bellows in miniature, or to an English pounce-box, some of which I have seen with flat sides considerably like them, but smaller. The middle part of an Icelandic snuff-box is made of wood, the neck and screw of brass.

neral method of making use of their chief luxury; for the more moderate snuff-takers will be satisfied by shaking some upon the back of their hand, and then inhaling it with their nostrils; or by expanding the fore finger and thumb, so as to form a little pit or hollow at the base of the thumb, which will contain half a nostril-full: but, by this method, more is wasted. It is, perhaps, one of the most disagreeable features among the generality of the Icelanders, both men and women, that their nostrils are always overflowing with this precious dust. The information which these men gave us was, that the governor of the island, Count Tramp, had just arrived in his ship, the Orion, from Denmark, and, that a man of war, from England, had but two days previous left Reikevig, where she had been staying some time, and had been entering into an agreement with the governor about permitting the island to trade with the English. In a few hours, we came within sight of the islands about Reikevig, which appeared to be pretty well clothed with grass, and to have on them both houses and cattle. Along the shore, also, were here and there scattered a few cottages, which, on account of their being covered with turf, were not easily distinguishable from the ground they stood upon, and, sometimes, only by the superior luxuriance of vegetation. Another boat was now seen coming from the shore, in which were Mr. Savigniac, an agent for Mr. Phelps, who had spent the winter there, and a Mr. Betreyers, a Danish Merchant, who could speak a little English. While these gentlemen were talking over commercial affairs below, I kept upon deck, watching, with my telescope, every little object as it came in view. The house of the physician, Doctor Clog (pronounced Clo), a neat white building, covered with boards, was pleasantly situated upon a flat grassy peninsula, and, a little beyond it, we discovered the small town of Reikevig. The most conspicuous feature in this town was a pretty large white building, roofed with boards, which, I concluded, was the residence of the governor, but was surprised on being told it was the work-house, or house of correction. On drawing nearer, however, it was not such a comfortable place as it appeared in the distance, and the houses in the town, which we had a good view of, as we came to an anchor in the harbor, exhibited a more favorable exterior. A long line of buildings, principally warehouses, and all made of wood, fronted the sea. The church was distinguished by its being of stone, and covered with tiles, and by having a small steeple, or little square wooden tower, for its two bells. On each side of these buildings, among the rocks, which on every side surround the town, were scattered miserable huts, but little raised above the level of the ground, although none of them are really formed under ground, nor, indeed, are any in the island so, as has been generally supposed. About three in the afternoon, we came to an anchor at a short distance from the town, close by the Orion, and, at four, we went on shore, landing upon a beach wholly formed of decomposed lava, of a black color, and, in some places, almost as fine as sand: here, a sort of moveable jetty, made of fir planks, was pushed a little way into the sea, that we might not wet ourselves, and, at least, a hundred natives, principally women, welcomed us to their island, and shouted on our landing. These good folks did not gaze on us with more pleasure, than we did upon them. It was now the season for drying fish, and they were employed in this operation at the time of our arrival. Some were turning those that were laid out to dry upon the shore; another groupe was carrying, in hand-barrows, the fish from the drying place to a spot higher up the beach, where other persons were employed in packing them in great stacks, and pressing them down with stones, to make them flat. Most of this business was performed by women, some of whom were very stout and lusty, but excessively filthy, and, as we passed the crowd, a strong and very rancid smell assailed our noses. The first peculiarity about the women, which strikes the attention of a stranger, is the remarkable tightness of their dress about the breast, where the jacket is, from their early infancy, always kept so closely laced, as to be quite flat, a practice which, while it must be a great inconvenience to themselves, entirely ruins their figure in the eyes of those who come from a more civilized part of the world. Their dress is not otherwise unbecoming, except that the waist is too long, and, from its warmth, it must be well suited to the coldness of this climate. Upon their heads, in their working, or common, dress, they wear a blue woollen cap, with a long point, which hangs down by the side of the head, and is terminated by a tassel, nearly resembling such as is worn by many of our horse-soldiers, in their undress uniform, and this tassel is often ornamented with silver wire. When they have this head-dress, their long and dirty hair is suffered to hang over their shoulders to a great length; but not so, when the *Faldur*, or dress-cap, is worn: then the hair is carefully tucked up, so that none of it is seen. As, however, I shall confine myself at present to the dress of those females whom I saw at work when I landed, I shall reserve my description of the turban, and of the costume of the richer people, till another opportunity. Over a great number of coarse woollen petticoats, which make them look of a most unnatural size, and a shirt of the same materials, they wear a thick petticoat, or rather gown without sleeves, (for there are two apertures for the arms,) made of blue or black cloth, and fastened down the breast, either by lacing, or, as is more common, with silver clasps*. A short jacket of the same, which has sometimes a little skirt, goes over this, and is fastened, likewise, about the breast with brass or silver clasps, or by lacing. Their stockings are of coarse wool, knitted and dyed black; and their shoes made of the skins of sheep or seals. Over the shoulders of many of them, on each side, were hanging thick ropes of horse-hair, coarsely braided, with a noose at the end, by which they carried the hand-barrows with fish. The dress of the men was pretty nearly the same as that of our pilots, except that their clothes were generally black, and their stockings, also. In laborious employments, both they and the women frequently threw off their jacket, and worked with nothing but their worsted shirt-sleeves over their arms. As to the features of this groupe of ladies, the generality of them were, assuredly, not cast

* This gown (*Upphlutur,* in Icelandic), however, is not, any more than the petticoats are, so long as to conceal much of their ill-shaped legs: otherwise, it would be a great hindrance to their walking among the rocks. I recollect one old lady, a constant laborer on the beach, who never had her dress come lower than her knees.

in nature's happiest mould, and some of the old women were the very ugliest mortals I had ever seen; but, among the younger ones, there were a few who would be reckoned pretty, even in England; and, in point of fairness of complexion, an Iceland girl, who has not been too much exposed to the inclemencies of the weather, will stand the comparison with ladies of any country. They are generally of a shorter stature than our women, but have a good deportment, and, to judge from their appearance, enjoy an excellent state of health. After having attentively surveyed this interesting assemblage, we repaired to Mr. Savigniac's house; but, as this was built in Norway, and not different from what a wooden house would be in our own country, it had no charms for me. I therefore hastened to take a ramble by the sea shore. A little rude bridge, formed of planks, across a streamlet, led me out of the town; and, passing two or three peasant's houses,*

* Close by these houses, and by all in the immediate vicinity of the sea, are contrivances for drying the fishing-dresses, which are made of untanned sheep skin, with the hair inwards, or rudely scraped off, and comprise the jacket and trowsers all in one piece,

I pursued my way among the rocks in search of plants. I cannot compare the country I here walked over, to any thing or place I know, which it so much resembles, as the summit of Ben Nevis; for, with the exception

giving the wearers a singularly wild and savage appearance. This dress is worn over their common clothes. The machines are of a simple structure; consisting of an upright stick, three or four feet high, and a smaller transverse bar, crossing this at the top, and turning on its centre: from this horizontal bar, hangs down at each extremity, a longer piece of wood, in such a manner as to form three sides of an oblong square. The annexed sketch conveys a sufficiently accurate idea of the whole. Two or three or more of these are placed near every fishing-house, so that, when the inhabitants return from fishing, with their wet dresses, they suspend them, by fitting them on the upper part of these machines, which turn about with the wind, in such a way that a current of air always passes through them.

of here and there a few patches of verdure, the whole was a mass of broken pieces of rock, not piled up in heaps, but forming a great plain, or, at most, only rising in a few hills, of a gentle and gradual ascent. Nearer the sea, some of these pieces of rock were covered with a little earth and grass, and in other places the interstices were frequently filled with *Trichostomum canescens*, among which grew many alpine plants, which again forcibly reminded me of the summit of our more elevated Scotch mountains, where the vegetation is by no means dissimilar. Among the most common lichens were *Endocarpon tephroides*, *Lecidea geographica*, a new *Lecidea* with a yellow granulated crust and brighter yellow shields, *Cetraria islandica* and *nivalis*, *Parmelia scrobiculata*, *fusco-lutea*, and *brunnea*, *Stereocaulon globiferum*, and *Bæomyces endivifolius*, and *vermicularis*. I met with but few mosses, except such as are extremely common almost every where. There was one, however, that approached, in habit, *Encalypta lanceolata*, a sketch of which I happen now to have by me, and from this, on comparison, it appears to have most affinity with *Dicranum latifolium*, but is probably different from both. *Buxbaumia foliosa* and *Catharinea hercynica*, were common on wetter grounds, and with them was an abundance of male specimens in fructification of Dr. Wahlenberg's *Catharinea glabrata*, which I did not distinguish from its neighbor of the same family till Mr. Bright the following summer brought home this plant with capsules, and I then recognised the new Lapland Moss I had often seen in Mr. Turner's *Herbarium*. *Lychnis alpina* was scarcely in flower; *Saxifraga tricuspidata*, *Fl. Scandin.* was in the same state. *Cardamine petræa*, *Draba incana*, and *contorta*, and a *Stellaria*, which appeared to agree with the description of *groenlandica*, were all plentiful. *Silene acaulis* and *Cerastium alpinum* were not yet in blossom. *Juncus trifidus* and *biglumis* were most abundant: the latter formed a considerable part of the herbage, intermixed with our more common grasses, and with *Festuca vivipara*. Late in the evening I returned to Reikevig, and slept for the last time on board the Margaret and Anne.

Thursday, June 22. This day was exceedingly cold and wet, and in the early part of it there was so thick a fog, that we could not see the town from our vessel. As soon as we had breakfasted, my luggage was conveyed on shore, and placed in Mr. Savigniac's house, where it was proposed, that, while we continued together, we should all meet at our meals; and where, with the addition of our ship-provisions to the good Icelandic mutton, fish, and scurvy-grass *(Rumex acetosa* and *digynus)*, we fared exceedingly well. I had this morning a favorable opportunity of looking at the town, which consists of about sixty or seventy houses, standing in two rows, of nearly equal length, at right angles with one another, so as to form the annexed figure, supposing the base of it to front the sea, and the upper part to run into the country. Those houses next the bay I have before mentioned, as being all built of wood: they face the north, and look, at a little distance, not unlike a number of granaries. The merchants' houses are built exactly like the warehouses; that is to say, of wooden planks, covered with the same materials; and are only to be distinguished by

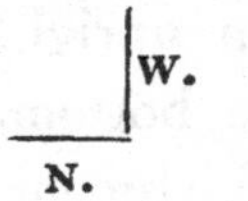

W.

N.

their having a few glass windows, and one or two wooden chimnies. These are all framed in Norway, then taken to pieces for stowage in the ship, and conveyed here. The warehouses are also shops, where the merchants retail cloth, earthenware, tin and iron utensils, sugar, coffee, tobacco, snuff, rye-flour, shoes, rum, in short, every necessary of life; and take, in exchange, for exportation, wool, tallow, fish, fish-oil, seal-oil, fox-skins, swan-skins, eider-down, worsted stockings, mittens, and, sometimes, dried mutton. At the western corner of this row of shops are the stocks, or, what might rather be called, a pillory; for the culprit stands upon a block, and has his arms fixed in two holes, formed by iron clasps, on the side of an upright pole, at about four feet from the bottom. From near this instrument of punishment, two rows of houses run parallel for some hundred yards, in a south direction, and form a tolerably wide street; but so encumbered with pieces of rock, that, if there were such a thing as a cart in the country, I fear it could not proceed half a dozen yards even up this, the high street of the capital. At the commencement of the right hand side, are two or three merchants' houses, and store-rooms; and, near them, is the residence of the learned Bishop of Iceland, Geir Videlin, or, as he is commonly called, Videlinus. His house differs in no respect from that of the merchants, except in being rather larger, and having more glass windows. Adjoining it, is the best house in the place (next to the governor's), which belongs to the *Landfogued;* it contains some comfortable rooms, and is well furnished. Still further up the street is a sort of tavern, where the Danes amuse themselves with cards, in a room which was built for the purpose of holding a considerable party, and was afterwards the scene of our Icelandic festivities. This building terminates the principal part of what forms the street: beyond it, are only a few cottages, made of turf; one of which was remarkable for its neatness, and for producing upon its roof and walls, besides a luxurious covering of grass, abundance of a *Draba,* which differed from the *contorta Fl. Scandin.* in having hairy capsules. It was here that I had my lodging, during the first part of my stay in Reikevig. The person of whom I hired it was of some consideration in the neighborhood; she being midwife to a very considerable district, with an income of twenty pounds a year from the Danish government, for which she had to furnish all her patients with proper medicine and attendance. As she had learnt her profession in Denmark, and had, moreover, been brought up, in the capacity of a servant, in the king's palace, at Copenhagen, she thought herself of more consequence than most ladies of her profession would do in any other country; and, although so much advanced in years, as to be nearer sixty than fifty, she was a constant visitor at the Iceland balls, and, at a reel, would dance the very fidler out of patience. This was almost the last house in the south-western angle. If two lines were drawn from the points of these two rows of houses, which I have just described, so as to form a square, it would, near the south-eastern corner, contain the governor's house, and, adjoining it, that of Mr. Savigniac; the former small, but, internally, well painted and furnished; and, not far from these, near the north side of the imaginary square, stands the cathedral, a considerable building, with large glass windows, which, however, as well as the tiles, are in a wretched state of repair; so much so, that the ravens, which abound in the country, are very troublesome during the time of service, by getting on the roof, and disturbing the congregation with their noise and dirt. Another building requires to be mentioned, situated almost by itself, on a large green, which occupies this part of the town, that is, the court of justice, where all causes are tried under the presidency of the *Tats-roed.* It is nothing but a large wooden building, with two or three good sized, but nearly unfurnished, rooms, which are, when not otherwise employed, in the occupation of the tailor of the place. Many of the houses in the town, as well as (though more rarely) those in the country, have small gardens attached to them, fenced in with high turf walls, and generally kept neat and free from weeds; but this latter circumstance arises, perhaps, more from the paucity of indigenous plants of any sort, and the tardiness of their growth, than from any particular industry of the inhabitants in destroying them. Cabbages, especially the rutabaga, turnips, and potatoes, with sometimes a few carrots, are attempted to be cultivated, but never arrive at any great degree of perfection. Probably, the best garden, both in point of soil and situation, in the town, was that of Mr. Savigniac; certainly, none was half so much attended to. Here we had, in the month of August, good turnips about the size of an apple, and potatoes as large as the common Dutch sort. Radishes and turnip-radishes were very good in July and August. Mustard and cresses grew rapidly and well. Mr. Phelps ordered some seeds of hemp and flax to be sown as soon as we landed; but, with all the care and attention that was given up to them, at the expiration of two months, the former had not reached to more than one foot high, nor had the latter exceeded six or eight inches: neither showed any appearance of flowering, but, on the contrary, both had ceased to grow, becoming materially injured by the frosts. I would not wish to be understood, that this garden is by any means a fair criterion to judge of the progress of vegetation in Iceland; for a more sheltered spot and richer soil were hardly to be met with. In other gardens, and especially out of the town, vegetation was extremely languid, and, even in the month of August, when the cabbages ought to be in their best state, I was in many gardens where a half-crown piece would have covered the whole of the plant, and where potatoes and turnips came to nothing. It must be remarked, however, that this was an extremely cold and wet season: in finer summers, with care and well-sheltered gardens, some of our more hardy vegetables may, doubtless, repay the natives for the labor of cultivating them*. On the outskirts of the town are

* It was not till after my return from Iceland, that I met with *Horrebow's Natural History of Iceland*, where I was somewhat surprised to find a chapter on the fruits of the earth; containing an account of the vegetables, which may be, and which are, produced there, differing extremely from what I have above stated. That author begins, by saying, "All kinds of things may be produced, fit for a kitchen-garden, and brought to proper maturity; (and, why not?) for this island is as proper for vegetation as Norway, having large plains and fields, and a great deal of good ground." I believe I need only mention, on the one hand, the total want of timber in Iceland, and, on the other, the immense forests which are met with in Norway, to convince any one that the former country is not so proper for vegetation as the latter.—" In the year 1749, when I came to Bessested, one of his majesty's palaces or seats, in Iceland, I found the garden in excellent order, and full of all kinds of vegetables,

a few scattered Iceland-built houses; but, with the exception of these, almost all the

fit for a kitchen: such as parsley, celery, thyme, marjoram, cabbages, parsnips, carrots, turnips, peas, beans, in short, all sorts of greens wanted in a family. I can vouch, with the greatest truth, that I never saw a garden with better things of the kind in it. They were all of good growth, and had all the properties that good garden-stuff ought to have. They were all in such plenty, that considerable parcels of them were dried and laid by for the winter, such as sugar-peas, and the like. I, myself, have taken up a turnip that weighed two pounds and a half. Hereby, I do not intimate that all were so big, but, only, that they are of a very good size. They have gooseberry-bushes, that produce fine and ripe berries."—I should be sorry to contradict any assertion of Mr. Horrebow's (who, in many respects, is entitled to considerable attention, and who appears to me to endeavor to separate truth from error, in several instances), to which he says, he was an eye witness; but this I must be allowed to say, that I never heard at all, in the island, of many of the vegetables which he mentions, as coming to such perfection; and, as to gooseberries, I have the authority of the *Tatsroed,* for stating, that they cannot be cultivated to the least advantage. Kerguelen, in confuting Mr. Horrebow's affirmation, that he ate currants from the garden at Bessested, inclines too much to the opposite extreme, when he says, "I believe it to be as difficult to raise turnips in Iceland, as pine-apples at Paris."

houses of Reikevig, are of Norwegian construction, and, indeed, principally inhabited by Danes; so that this cannot properly be called an Icelandic town: nor is there such a thing in the whole country; for, depending, as the natives must do, almost entirely upon the scanty produce of their own island, and requiring a considerable tract of country for the maintenance of a few half-starved sheep, such societies, as would form a town, or even a village, would be highly prejudicial and unnecessary. There are merchants, who reside in other parts of the coast; but by far the greatest number of Icelanders bring their produce to this place; some coming from the most northern and eastern parts. Iron is what they are most anxious to procure, for their horses shoes, their scythes, and implements for cutting turf and digging. Those who live in the interior of the country, and have no opportunity of going down to the coast in the fishing season, take back, in exchange for their tallow and skins, the dried heads of the cod-fish, and such of the fish themselves, as are injured by the rain, and not fit for exportation. These form the principal article of their food, and are eaten raw, with the addition of butter, which, after the whey has been expressed, is packed down in chests, and kept for several years. Their drink is either water, or sour milk, or whey, and sometimes, but rarely, new milk from their cows or ewes. *Skiur*, which is thick curd, may also be reckoned a common article of food: this they prefer after it has acquired a sour, and even a rancid, taste; though, when fresh, or when it has attained only a slight degree of acidity, and is eaten with cream and sugar, it is really an enviable article of luxury. The country immediately about Reikevig, and, indeed, for twenty or thirty miles from it, is ugly, barren, and scarcely to be called hilly. An extensive fresh-water lake comes close up to the back part of the town, but is on every other side, except that nearest the town, surrounded by bog, with here and there a piece of rock interspersed. Not a tree or shrub is any where to be seen, and all attempts that have been made in the most sheltered parts of the place to cultivate firs and other hardy trees, have universally failed, as have those which have been made for the cultivation of corn. This lake empties itself into the sea by a small stream which runs by the side of the town, in a course of not more than a few hundred yards. Towards the east side of the lake, on a gentle elevation, where a tolerably rich herbage is produced, a prodigious number of great pieces of rock are scattered about, in the utmost disorder: some of them are of vast size, three or four times the height of a man, and about as wide as they are high; yet there is no mountain in the neighborhood from which they could have rolled; nor could I find any cavities near the place on which they stood, that would render it probable they were thrown up by an earthquake; neither do they appear, just in that spot, to have undergone the operation of fire, although some rocks, close by, have evidently been in a state of fusion. On the shore, in several places near the town, are many rudely-formed basaltic columns, standing close together, in a perpendicular direction, some from one to two and three feet in diameter: they are obscurely angular, and, on the top, are generally either concave or convex. They appeared to me exactly of the same nature as those of Staffa, and are found, also, on many of the islands near Reikevig. Being anxious to visit the boiling spring, about two miles and a half to the eastward of Reikevig, the steam from which was pointed out to me from a little eminence near the town, I set out about one o'clock for that purpose; but, after getting enveloped in a labyrinth of bogs during a heavy rain, I was obliged to return without being able to reach it, and with but a few plants, which I had not found the preceding day. This, however, was not to be wondered at, since the most part of the tract I went over was either barren rock, or a morass, where the grasses showed no appearance of coming into flower. Near the shore, I saw several different sorts of the duck tribe, and, especially, a number of the eider-fowl. Cormorants were abundant. Cast upon the beach, were scarcely any but the more common sea-weeds of Scotland, as *Fucus palmatus, esculentus, digitatus, ciliatus, dentatus, purpurascens, saccharinus*, and a variety of the latter, with a twisted frond, *plumosus, flagelliformis, rubens*, and *Conferva fœniculacea* of Hudson. *Fucus ramentaceus*, which has hitherto been found no where but in Iceland, was the only rare species, and this was here in great plenty. Some of these were growing in the basins among the rocks. Of shells there were very few. I remarked a large *Balanus*, which seemed to me new. It is well figured in Povelsen and Olafsen's Voyage, plate 14, but I cannot, any where, find a description of it. *Mya truncata*, *Venus islandica*, and a beautiful, but to me unknown, species of *Lepas*, a *Bulla*, and a few *Turbines*, were the only other shells I met with. Land-birds are extremely rare. All that I saw in this walk were Ravens, the Snow Bunting (here called *Snoe-fugle*), which has rather a pleasant note, not much unlike the Linnet's, but more interrupted, Snipes, and the common Wagtail.

Friday, June 23. Another day of rain kept me almost entirely confined to the town. In the morning, accompanied by Mr. Jorgensen, I made a visit to the Bishop, Geir Videlin, or, as he is commonly called, Videlinus. He has a good library; indeed, very much better than I expected to have seen in Iceland: it appeared to contain five or six hundred volumes, among which are several Dutch editions of the Classics, a perfect, but uncolored, copy of the *Flora Danica*, and a fine folio edition of an Icelandic Bible, printed in the island, in 1584, which has a curious and well-executed frontispiece, cut in wood, by the hands of Bishop Guthrandr Thorlaksen, without any other instrument than a penknife: the same person, also, set the letter-press. Bishop Videlinus has, besides, a very beautiful Icelandic manuscript, written in the year 1525, in defence of the Christian Religion. Till within a few years, the residence of the bishops (for there were two) was at Skalholt, but it was found more convenient to have the see removed to the principal place of resort and traffic, so that the clergy have now the opportunity of transacting business with the bishop and the merchant at the same time. When they come, they take up their abode with the bishop, who, on this account, can hardly live upon his salary of fifteen hundred dollars a year, which is all that is allowed him by the Danish government. He is a stout and handsome man, and wears black clothes, with half-boots. His hair is remarkable for being almost white, though not from age, as he is not more than forty-five. Both he and his lady are native Icelanders: the latter dresses in the true Icelandic fashion, and, indeed, her costume of ceremony is extremely rich and handsome. The bishop's library is almost continually filled with visitors, it being the principal place of resort for those who are desirous of studying, and almost the only one that affords them the advantage of a good collection of books: among other men of learning, I used frequently to meet here Finnur Magnusen *, a man highly celebrated among the modern Icelanders for his abilities as a poet, as well as for the variety and extent of his attainments as a scholar. To him I was indebted for a present of many Icelandic books, one of which was sufficiently remarkable in having for its title, *The Georgics of Iceland* †! It is considered a

* In the former edition of my Tour, this gentleman has been erroneously called Magnus Finnusen. The kindness of my Icelandic friend, Mr. Sivertsen, has enabled me to make this and other similar corrections.

† My ignorance of the Icelandic language rendered me, unfortunately, unable to read this book, which must have been a matter of considerable curiosity, unless, indeed, it was altogether fictitious; as the Icelanders have no husbandry whatever to employ them,

scarce book, and a fine poem; though, as the Etatsroed told me, many of the rural occupations spoken of in it are by no means applicable to the country it professes to describe. If I mistake not, it was written by one Povelsen*, an ancestor of Finnur Magnusen, and this latter had himself translated it into

or to be sung about, except the care of their cattle. The author of this work, which, by the bye, is but a small one, could not begin with the words of Virgil,

"Quid faciat lætas segetes, quo sidere terram
"Vertere, Mæcenas, ulmisque adjungere vites
"Conveniat, quæ cura boum, qui cultus habendo
"Sit pecori? apibus quanta experientia parcis:
"Hinc canere incipiam."

The oxen and the flocks are all he could have found in Iceland: the corn and the vines assuredly do not exist there, and even the acuteness of my friend, Mr. Kirby, would have been puzzled to have found one of his two hundred and twenty-two species of British bees in the island.

* In this instance, as, I am afraid, in some others relative to the names of persons and places, my memory has not served me faithfully; for I find, by Dr. Holland's valuable Dissertation, just published, *On the present State of Education and Literature in Iceland*, that the name of this author, instead of Povelsen, is Eggert Olafson.

Danish verse. As a proof of the talents and readiness of this young man, it may not be amiss in this place to mention, that though, at the time of our arrival, he did not know a word of English, yet he made so rapid a proficiency in the language, that, during the stay of the Talbot sloop of war, only two months after, he submitted a copy of English verses to one of the officers of that vessel for his correction. An exceedingly long complimentary ode *, also, in Icelandic poetry, was presented by him to Captain Jones of the Talbot, with a latin translation by the side.

Saturday, June 24. To-day the captain of our vessel and Mr. Savigniac accompanied me to the little island of Akaroe, situated in the bay, at a short distance from the town of Reikevig, for the purpose of seeing the eider-ducks, which breed on this, as well as on all the other uninhabited islands, in great quantities. It was a windy day, and we had a rough passage in a small Icelandic boat, over which the waves were continually

* See Appendix D.

beating. These boats, which are rowed by two men, are very high, both at the head and stern, and, by being made sharp as well fore as aft, are capable of being rowed with equal facility both ways: the larger ones, however, have a rudder. The sides of the boats, instead of bellying out, like ours, are nearly flat, and applied to each other at acute angles, so that a transverse section would appear almost like the letter V: at the same time they are so deep, that they require to be supported by a very considerable quantity of water to keep them afloat, and as often as this is not the case they necessarily fall down on their sides, which renders the getting in and out a matter of some nicety. They are, nevertheless, safe boats, and accidents are seldom heard of from their oversetting. On our landing on the rocky island, we found the eider-fowls sitting upon their nests, which were rudely formed of their own down, and generally built among the old and half decayed sea-weed, that the storms had cast high up on the beach, but sometimes only upon the bare rocks. It was difficult to make these birds leave their stations: indeed, so little inclined were many of them to do it, that they even permitted us to handle them whilst they were sitting, without their appearing to be at all alarmed. Under each of them were two or four eggs: the latter is the number they lay, but in many instances the birds had been robbed of half, which had been taken for food by the natives, who prefer those that have young ones in them. These eggs, which are in Iceland esteemed a delicacy, though in England they would not be considered equal to those of our barn-door poultry, are of a pale olive-green color, and rather larger than those of a common duck. In one part of the island, where there was a considerable quantity of rich loose mould, the Puffins * breed in vast numbers, forming holes three or four feet below the surface, resembling rabbits' burrows, at the bottom of which

* *Alca arctica Linn.* called in Iceland *Soe-papagoie* and *Præst,* in Cornwall and in the south of Scotland, according to Mr. Neill, *Pope.* In Kamtschatka and the Kurilschi Islands, the inhabitants wear the bills of these birds about their necks, fastened to straps; and, according to the superstition of those people, their *Shaman* or priest must put them on with a proper ceremony, in order to procure good fortune. See *Latham's General Synopsis of Birds,* vol. v. p. 317.

they lay a single white egg, about the size of that of a Lapwing, upon the bare earth. Our people dug out about twenty of these birds, which they afterwards assured me made an excellent sea-pie. The Icelandic fishermen catch the Puffins, and use their flesh for bait; being persuaded that the cod prefer it to any thing else. On all the rocks about this island, which were covered at high water by the sea, was growing in considerable quantity the *Fucus palmatus* of Linnæus, known by the natives under the name of *Sol**. As an esculent *Fucus*, this species seems to be preferred to all others, at least in northern countries. On the Scotch coasts, it is eaten raw by the natives, and, in the county of Caithness in particular, I have seen a number of women and children gathering it from the rocks, and making a meal of it, devouring it with avidity. In Iceland, also, it is very commonly eaten, but

* According to Povelsen and Olafsen, *Sol* is a considerable article of trade with the inhabitants of the town of Oreback, who receive in exchange for it butter, meat, cattle, and wool. A *Voet* (about eighty pounds weight) of this *Fucus*, when dried, sells there for seventy fish, at two skillings a fish, or five shillings and tenpence English.

seldom while it is fresh. It is generally well washed in clean water, and exposed upon the rocks, or on the ground, to dry, when it gives out a whitish powdery substance, which covers the whole plant, and is sweet and agreeable to the palate. It is then packed down in casks, to keep it from the air, and is preserved in this state ready to be used, either raw with fish and butter, or boiled down in milk to a thick consistency, as is more common with people of property, who mix with it, if it can be afforded, a little flour of rye. This species is the true *Alga saccharifera* of Biarne Povelsen, who has written a dissertation upon it. It has been, however, the opinion of many Fucologists, that the *Sol* of the Icelanders is the *F. saccharinus* of Linnæus; misled, probably, by the name of the latter, which, however, does not give out a *saccharine* powder, but merely saline particles, by no means agreeable to the taste. Of this, Gmelin, in his *Historia Fucorum*, page 198, says, "certumque quoque est, saccharum, quod, profert, non nisi salem marinum esse, in substantia Fuci efflorescentem, qui propterea levitèr gustatus dulcedinis sensum linguæ imprimit, quique purgantem effectum edit, si Fuci ingesta copia nimia fuerit, sale tum fibras intestinales vellicante." The learned Etatsroed of Iceland has written a full account of the three esculent *Fuci* of his country, *F. palmatus, F. digitatus,* and *F. esculentus,* which was printed at Copenhagen last spring. Of this work he very kindly presented me with a copy for myself, and also one for Mr. Turner, with whose *Historia Fucorum* (as far at least as was then published) he was not unacquainted. The number of quotations from various authors in the Etatsroed's little work was a sufficient proof of his having paid great attention to the subject on which he wrote, and of his possessing botanical books, which a stranger would little expect to meet with in Iceland. I much regret the loss of these two pamphlets, as they contained, not only a complete account of the mode of preparing the *Fuci* for food, but also a very accurate representation of the three species, from drawings (if I mistake not) made by the Etatsroed himself.

Sunday, June 25. This morning, I visited the more elevated parts of the country about Reikevig, and found them composed wholly of broken, and generally small, pieces of rock, for the most part perfectly barren; though in places, here and there, were some patches of vegetation, among which I met with a few interesting plants. *Vaccinium uliginosum* was abundant, and its charming blossoms delighted me much, the more so as I had never previously seen it in perfection. *Dryas octopetala,* of which the inhabitants gather the leaves and make a sort of tea of them was every where extremely common, but hardly yet in flower, and the same was the case with *Lychnis alpina.* A remarkably woolly-leaved *Salix,* which I took for *lanata,* and two or three other species, of stunted growth, were the only plants that elevated themselves to the height of even five or six inches from the ground. *Saxifraga (tricuspidata? Fl. Scandin.)* grew plentifully among the rocks; which also produced *Splachnum vasculosum* and *mnioides,* though sparingly; but I was most pleased with a fine new species of *Cornicularia,* allied to *C. bicolor,* but three or four times as large, and all over of a grey color. I met with only one patch of it, intermixed with *Trichostomum canescens*, in a rocky situation. From these hills, though at a considerable distance, I could perceive the steam from the hot spring, and, taking a different route from what I had done when I made a former attempt, I at length, with some difficulty, arrived at it. While yet full a mile from the spot, the superior verdure of the grass, that was within the influence of the heat, was very remarkable. What struck me as most extraordinary in this spring, though I afterwards found it not to be uncommon in Iceland, was the circumstance of its being actually situated in the middle of a cold stream, bubbling up from some little cavities, which were formed in a whitish siliceous incrustation, that covered a considerable portion of the bed of the river, and extended on one side of it, even as far the shore, where its surface was covered with numerous minute mammillæ. This incrustation is a deposit from the water, and the mammillæ are probably caused by the irregular falling of the water upon it in drops. On dipping in the water my little pocket thermometer, which was graduated to no more than 120° of Fahrenheit's scale, but was the only one I had with me at the time, the quicksilver instantly rose to the top of the tube. I found lying dead in the hot water a number of eels*, not more than four or five inches long: these had, doubtless, been conveyed down by the rapidity of the current to the heated part of the water, which, as it affects the whole width of the stream, must be an effectual barrier to the migration of fish, and of other aquatic animals. I remarked, however, no others in this water, except one or two specimens of a *Dyticus,* which I was not able to catch, but which appeared to be the same as our *D. acuductus.* Almost in the hottest part of the water, I gathered *Conferva spiralis Dillw.;* but it had lost all its color, and had probably only floated into that situation, not being really a native of it: a species, also, which appeared to me to be new, grew attached to the banks, at a very

* Povelsen and Olafsen have mentioned the circumstance of small eels being found dead near the heated waters of Iceland, and remark that, although large eels are known to exist in the river, they have never been met with lying dead, as the smaller ones.

short distance from the bubbling water: it was most nearly allied to Dillwyn's *Conf. dissiliens.* *Conf. vaginata Dillw.* flourished in great perfection on a bank of earth, which rose immediately from the heated water, where it was constantly exposed to the steam. In the same situation, and equally vigorous, were *Gymnostomum fasciculare, Fissidens hypnoides,* and *Jungermannia angulosa,* all, except the last, bearing ripe capsules. On my return, I saw plenty of Snipes, in the boggy places, and, among the rocks, an arctic fox * *(Canis Lagopus L.)* which was changing its white winter dress for a summer one, being partly white and partly grey. These animals are extremely numerous in

* The dusky appearance of this animal, which I had always supposed was only the summer coat, (or that of a young fox which had not reached its second winter,) I find, is noticed by Mr. Pennant, in his *Arctic Zoology,* and considered as making a distinct species, though for what reason I do not see; since he himself observes that the color of the young fox is dusky. Kerguelen says there are black, blue, red, and white foxes in Iceland. It may not be improper to observe, in this place, that I never saw the common fox in Iceland, nor heard, from any of the natives, of its being found there.

this country, living upon the Ptarmigans and their eggs, as well as upon young lambs: their fur is thick, but too short for muffs and tippets of the present day, although in some of the grey-colored ones it is exceedingly fine and beautiful. They are sold in Reikevig for about one shilling and sixpence a skin. This fox is probably not to be considered as originally a native of Iceland; for the Icelanders have a tradition*, that one of the ancient kings of Norway, to punish the inhabitants for their disaffection to the mother country, sent over some foxes to the island, where they have rapidly increased, to the great injury of the flocks. The few rats and mice†, that are said to exist here, are

* This tradition, in all probability, rests upon no better authority than one which is prevalent in Ireland, that the breed of magpies, which now infest that island to such a degree, as to be highly injurious, was originally imported by the English to plague them. It is more likely, if the Icelandic foxes be not really natives of the country, that they found their way thither from the neighboring coast of Greenland on the floating masses of ice.

† Speaking of the native animals of Iceland, Pennant, in his *Arctic Zoology, Introduction*, page lxx. suspects, "that there is a species allied, as Doctor

brought by ships from other countries. Indeed it appears, that the truly indigenous animals of the class, *Mammalia*, are reduced to the small number of amphibious ones,

Pallas imagines, to the *Œconomic Mouse*; for, like that, it lays in a great magazine of berries, by way of winter stores. This species is particularly plentiful in the wood of Husafels. In a country where berries are but thinly dispersed, these little animals are obliged to cross rivers to make their distant forages. In their return with the booty to the magazines, they are obliged to repass the stream; of which Mr. Olafsen gives the following account: 'The party, which consists of from six to ten, select a flat piece of dried cow-dung, on which they place the berries on a heap in the middle; then, by their united force, bring it to the water's edge, and, after launching it, embark and place themselves round the heap, with their heads joined over it, and their backs to the water, their tails pendent in the stream, serving the purpose of rudders.' When I consider the wonderful sagacity of beavers," continues Mr. Pennant, "and think of the management of the squirrels, which, in cases of similar necessity, make a piece of bark their boat, and their tail the sail, I no longer hesitate to credit the relation." I am sorry such a ridiculous story should have been believed by a British zoologist. Iceland certainly produces no species of *Mus* which our country does not possess, and the mice that are found there are not likely to be furnished with any instinct or faculties superior to those of our own mice. The circumstance related

which are found on the shores. The white bear is now and then conveyed to the northern coasts, by the floating ice-islands, from the opposite shore of Greenland, but none had been over since the preceding year, and those were soon dispatched by the people living in the vicinity, who are with reason afraid of so formidable a neighbor establishing himself among them. Their skins are always the property of the king of Denmark. * As I entered the town of Reikevig, on my return in the afternoon, I was surprised to find a guard of twelve of our ship's crew, armed with muskets and cutlasses, standing before the governor's residence, and still more so, when, shortly after, I saw the governor himself, Count Tramp, come out of his house, as a prisoner to Captain Liston, who, armed with a drawn cutlass, marched

above, is laughed at by the more sensible Icelanders, and the species that performs these extraordinary feats, which, according to Povelsen, is the *Mus sylvaticus* of Linnæus, is not, to my knowledge, found in that country.

* For a detailed account of these transactions, which ended in a complete revolution, see Appendix, A. and B.

before him, and was followed by the twelve sailors, who conducted the Count on board the Margaret and Anne. At the same time I also observed the British colors flying over the Danish, on board the Count's ship, the Orion, which, I subsequently learned, had been previously made a prize to our English letter of marque. I had all along observed a great dislike on the part of our countrymen to the governor: this, as well as the apparent acts of violence, that had just been committed, was caused by information, which Mr. Phelps had received, from what might have been supposed good authority, that Count Tramp had been using his influence to prohibit the trade with the English, contrary to the articles of an agreement, entered into, by him and the captain of an English sloop of war, that had been in Reikevig harbor just before our arrival. During this transaction, the inhabitants of the town, most of whom were witnesses to it, offered no resistance, but looked on with the most perfect indifference. Many of them were idling about the town (it being Sunday), armed with their long poles, spiked at one end with iron, which they use for the purpose of assisting them in walking over the frozen snow, and half a score of the lustiest of these fellows might with ease have overpowered our sailors, who were almost as wholly unacquainted with the use of firearms as the Icelanders, and, were, moreover, a most wretched set, picked up from the vilest parts of Gravesend. In the evening, the bishop waited on Mr. Phelps, and entreated that the Count might be allowed to have his liberty, or, at least, that permission might be granted for him to remain on shore as a prisoner. Both these requests being refused, he begged that he himself might be suffered to go on board, and speak to him; but being disappointed in this third request, also, he came to me, and, after expressing the pleasure he felt on the information he had received, that my object in visiting Iceland was of a peaceable nature, as a naturalist, adding every now and then, with much emphasis and feeling, "*tibi semper pax est,*" he hoped that I would use my influence with Mr. Phelps, at least to permit the governor to come on shore for a few hours; at the same time offering, as a surety for his returning to the ship, that his own son, who was then standing in tears by his side, should be sent on board, during the Count's absence. It was thought proper, however, not to grant this wish. We witnessed a more affecting scene, soon after, when the Count's secretary, a most amiable young man, about seventeen years of age, a native of Norway, came and pleaded strongly for the release of his master; begging, if that could not be complied with, that he himself might be allowed to go on board, and remain with him in his confinement. When the latter was acceded to, he dried his tears, and, after expressing his gratitude for the permission, hastened to convey his bedding, &c., together with those of the Count on board the ship.

Monday, June 26. After the preceding day's transactions, it was thought possible that some disturbance might be raised, either by the Danes residing in the town, or by the natives; but all was quiet, and, to prevent any effectual opposition on their parts, the arms of the inhabitants were secured, which did not amount in the whole to above twenty wretched muskets, most of them were quite in a useless state, and a few rusty cutlasses. An incessant and heavy rain, till about six or seven in the evening, prevented my botanising; but, as we had no darkness, even at the hour of midnight, I could just as well pursue my employment then as in the middle of the day. The unpleasant light, caused by the horizontal rays of the sun striking on the ground, so beautifully described * by Linnæus, when botanising in Lapland, is not experienced here; for the sun, in this part of Iceland, is never altogether above the horizon at midnight, nor, if it had been so, would it have had that effect this summer, there being no one period, that I recollect, during the continuance of the longest days, when the horizon in the north was perfectly free from clouds. At such times as the sky is not altogether overcast, the light at mid-

* "Fugit me quid sit, quod visum in alpibus nostris, tempore nocturno, ita confundit, ut non tantâ claritate possimus objecta distinguere ac mediâ die, licet sol æque clarus exsistat; sol enim horizonti proximus radios horizontales dispergens vix pileo ab oculis abigi potest; umbræ dein herbarum extenduntur in infinitum et implicantur inter se, tremunt deinde spirante aquilone, ut vix videre et distinguere queamus objecta diversissima."—*Linnæi Fl. Lapp. edit. 2da. p.* 137.

night, at this season, is about as great as that of a moderately dull noon in winter in England. In a walk of a few miles to the south of the town this evening, I met with ***Rubus saxatilis*** (sparingly in flower), ***Polypodium arvonicum***, plentiful, ***Trichostomum ellipticum***, and ***Hypnum filamentosum***, growing among the rocks. In bogs I found two new species of ***Carex***, and ***Meesia dealbata***, with fully-formed capsules.—This evening Mr. Jorgensen took possession of the governor's house, and removed his residence thither; but I do not recollect, exactly, whether it was from this period, or, as I rather think, shortly after, that he was considered as governor of Iceland.

Tuesday, June 27. This being the day appointed for paying our respects to the old *Stiftsamptman* *, Olaf Stephensen, who has the

* *Stiftsamptman* is the Icelandic title for the governor, and, consequently, belongs to Count Tramp. But, as this gentleman (Stephensen), on account of his services to the country, was allowed to retire from his government, and still retain the title of *Stiftsamptman*, I shall, by way of distinction, apply it in this work to him, and, in mentioning Count Tramp, shall use the term of Governor.

title of *Geheime Etatsroed,* and was formerly governor of the island, Mr. Phelps, Mr. Jorgensen, and myself embarked at twelve o'clock in an Icelandic sailing-boat, with eight rowers, and had a passage of about four miles to his house, which is on the pleasant little island of Vidöe. As we approached this island, we got a view of the house, standing in a well-sheltered situation, between two hills, and, at a little distance, it had the appearance of a very respectable residence, being larger, and with something more imposing in its exterior, than any other I had yet seen. It is built of stone cemented together, has a number of glass windows, and is covered with a boarded roof. However, when we landed and came nearer, we perceived à lamentable want, as well of carpenters and bricklayers, as of glaziers. The glass, in such of the windows as still preserved any, was of the most ordinary quality; and in most instances the panes were broken, though this was concealed from view on the outside by a wooden shutter. The door-way was in the centre, but hid by a miserable sort of wooden porch, on each side of which was a door for entrance, which, if kept in better repair, might answer very well for a screen to the cold winds, but, in its present shattered state, is neither fit for use nor ornament. I could compare it to nothing so much as a pig-stye placed against the wall, and made rather higher than usual. However, with all this want of artificers, which appeared on the exterior of the house, there was a possessor within, whose reception of us and whose excellent fare would do credit to the actual governor, much more to the ex-governor, of any country, and deserve to be mentioned rather particularly. Indeed, I am the more tempted to enlarge upon this subject, as it is the first opportunity I have had of seeing the manners of a well-bred Icelander, and it is scarcely possible to have a more favorable one. When we were within a few hundred yards of the house, the Stiftsamptman came to welcome us to the country, and to his little island. He had a fine and healthy countenance, and, although in his seventy-eighth year, had the perfect use of his faculties. In conversation he was extremely fluent and animated. He wore, on this occasion, his full-dress uniform as Governor of Iceland, except the sword. His coat was of scarlet cloth, turned up with green, and ornamented with gold lace: his pantaloons of blue cloth, with gold trimmings; and he had half-boots with gold bindings and tassels, and a three-cornered hat, likewise ornamented with gold tassels, and trimmings of the same, and with a long white feather. We were immediately ushered through the portico, where we were obliged to stoop at the door-way, into a spacious hall, with a large wooden staircase; and hence through a large and lofty parlor into his bed-room, where I presented to him a letter of introduction, and a present of prints and books from Sir Joseph Banks, whose very name made him almost shed tears. During the time that Mr. Jorgensen was translating the letter to him, he frequently interrupted his reading, to relate some of the many noble and generous acts which Baron Banks (as he called him) had done for his countrymen. He asked a hundred questions about him in the most affectionate manner, particularly respecting his age and health. Then he related anecdotes of what passed during Sir Joseph's stay in the island thirty-seven years ago, in a manner which at once convinced us of the excellence of his memory, and of his gratitude to, and high esteem for, the great benefactor of Iceland. He told us of his liberal presents, of the splendor with which he travelled, and of the many Icelanders, who, having during the present war with Denmark been made prisoners by the English, had been released, and supplied with money till their return to their country, by Sir Joseph Banks' generosity. London, he observed, might produce as good a man, but it could not produce a better. When we asked him to return to England with us, he said, he would, if he were but ten years younger, were it only to see Baron Banks. He was delighted with the presents from Sir Joseph, and especially with some beautiful engravings of the Geysers, taken from drawings made by Sir John Stanley, in 1789. This gentleman, also, the Stiftsamptman frequently mentioned, and I was vexed that my not having the honor of being acquainted with him prevented my answering the various questions, that were put to me respecting him. During our conversation, some rum and Norway biscuit were offered us, and we then took a little walk about the island, which is scarcely more than two miles in circumference, and is one of the most fertile spots belonging to Iceland, producing some of the best sheep, besides excellent cows, horses, peat, and good water. We were shown with great pleasure the immense number of eider-ducks which breed on Vidöe, and which were now sitting on eggs or young ones, exhibiting a most interesting scene. The Stiftsamptman made us go and coax some of the old birds, who did not on that account disturb themselves. Almost every little hollow place, between the rocks, is occupied with the nests of these fowls, which are so numerous, that we were obliged to walk with the greatest caution, to avoid trampling upon them; but, besides this, the Stiftsamptman has a number of holes cut in the smooth and sloping side of a hill, in two rows, and, in every one of these, also, there is a nest. No Norfolk housewife is half so solicitous after her poultry, as the Stiftsamptman after his eider-ducks, which, by their down and eggs, afford him a considerable revenue; since the former sells for three rix-dollars (twelve shillings) a pound. It is collected from the nests, which the ducks line, or rather form, with it, to afford their young a warmer and more congenial situation, stripping for the purpose their own breasts of a covering which nature has kindly given at this season. When taken away, the old bird replaces it, and, according to Mr. Pennant, this is occasionally done as often as three times, the drake supplying the deficiency in case the down of the duck is completely exhausted. Cats and dogs are, at this season of the year, all banished from the island, so that nothing may disturb these birds. It one year happened that a fox got over upon the ice, and caused great alarm: it was long before he was taken, which was at last, however, though with difficulty, effected, by bringing another fox to the island, and fastening it by a string near the haunt of the former, by which means he was allured within shot of the hunter. Such an island as Vidöe is well bestowed on the present owner, by the Danish government, for the services done to his country, during the fifty years that he was in office. It is considered worth one hundred dollars (twenty pounds) a-year, in addition to which, the full pension of fifteen hundred dollars is continued to him, as if he were still actual governor; nor is it as a magistrate alone that this gentleman is deserving of the greatest praise, but also as a man of science. His researches into the history of his own country, and his valuable communications on various subjects relating to it, which have been sent to Copenhagen, have gained him many honorary marks of distinction from different learned societies, and those, not merely of Denmark and Norway, but also of other nations. It has seldom, if ever, fallen to my lot to see, even in those places which are most distinguished for the cultivation of science, so large a collection of diplomas and honorary medals, as in this remote corner of one of the most remote countries of Europe. I met with no plants upon the island, that I had not seen in the neighborhood of Reikevig, except *Erigeron alpinum*, which, however, was not in flower. We had scarcely reached the extremity of our walk, when a servant came to announce that dinner was on the table: consequently, we were obliged to return, though rather against our inclinations; for the earliness of the hour, it not being more than half past one, and our having already taken some refreshment, had kept us from being hungry. We found the table set out in the large room which I have already mentioned. It had a tolerably good boarded floor, and walls that once were white-washed. The furniture consisted of five wainscot chairs, a table, and two large chests of drawers, on which were displayed such articles of use as approached the nearest to china; some of them, I believe, really were so. Two closet doors were also opened, and exhibited a considerable quantity of excellent silver plate. Two large and old-fashioned mirrors occupied the space between the windows, and beneath them were marble slabs, placed upon gilded feet; but they were broken, and lay completely out of a horizontal direction. About sixty prints and drawings, some of them in frames, and a few glazed, concealed in some measure the nakedness of the walls: they were, it must be confessed, for the most part, of a very ordinary stamp; but, as many of them were portraits of the Stiftsamptman's friends, or prints of the sovereigns, and other great men of Denmark, they had their value, and their names and titles were detailed to us with evident satisfaction. Such as it was, it might truly be said to be the best collection of prints and pictures in the country. When we sat down to table, a little interruption was caused by the breaking down of the chair upon which his Excellency had seated himself; but this was soon settled, as there fortunately was still a vacant one in the room to replace it. The arranging of a dinner-table is attended in Iceland with little trouble, and would afford no scope for the display of the elegant abilities of an experienced English house-keeper. On the cloth was nothing but a plate, a knife and fork, a wine glass, and a bottle of claret, for each guest, except that in the middle stood a large and handsome glass-castor of sugar, with a magnificent silver top. The natives are not in the habit of drinking malt liquor or water, nor is it customary to eat salt with their meals. The dishes are brought in singly: our first was a large turenne of soup, which is a favorite addition to the dinners of the richer people, and is made of sago, claret, and raisins, boiled so as to become almost a mucilage. We were helped to two soup-plates full of this, which we ate without knowing if any thing more was to come. No sooner, however, was the soup removed, than two large salmon, boiled and cut in slices, were brought on, and, with them, melted butter, looking like oil, mixed with vinegar and pepper: this, likewise, was very good, and we with some difficulty cleared our plates, earnestly hoping we had finished our dinners. Not so; for there was then introduced a turenne full of the eggs of the Cree, or great tern, boiled hard, of which a dozen were put upon each of our plates; and, for sauce, we had a large basin of cream, mixed with sugar, in which were four spoons, so that we all ate out of the same bowl, placed in the middle of the table. We petitioned hard to be excused from eating the whole of the eggs upon our plates, but we petitioned in vain. "You are my guests," said he, "and this is the first time you have done me the honor of a visit, therefore, you must do as *I* would have you; in future, when you come to see me, you may do as *you* like." In his own excuse, he * pleaded

* In Kamtschatka, acccording to Kracheninnikow, when a feast is given to a person for the purpose of

his age for not following our example, to which we could make no reply. We devoured with difficulty our eggs and cream; but had no sooner dismissed our plates, than half a sheep, well roasted, came on, with a mess of sorrel *(Rumex acetosa)*, called by the Danes scurvy-grass, boiled, meshed, and sweetened with sugar. It was to no purpose we assured our host that we had already eaten more than would do us good: he filled our plates with the mutton and sauce, and made us get through it as well as we could; although any one of the dishes, of which we had before partaken, was sufficient for the dinner of a moderate man. However, even this was not all; for a large dish of *Waffels*, as they are here called, that is to say, a sort of pancake, made of wheat-flour, flat, and roasted in a mould, which forms a number of squares on the top, succeeded the mutton. They were not more than half an inch thick, and about the size of an octavo book. The Stiftsamptman said he would

gaining his friendship, the master of the house eats nothing during the repast; "Il a la liberté de sortir de la Jourte quand il le veut; mais le Convié ne le peut qu après qu'il s'est avoué vaincu."

be satisfied if each of us would eat two of them, and, with these moderate terms, we were forced to comply. For bread, Norway biscuit and loaves made of rye, were served up; for our drink, we had nothing but claret, of which we were all compelled to empty the bottle that stood by us, and this, too, out of tumblers, rather than wine glasses. It is not the custom in this country to sit after dinner over the wine, but we had, instead of it, to drink just as much coffee as our host thought proper to give us. The coffee was certainly extremely good, and, we trusted it would terminate the feast; but all was not yet over; for a huge bowl of rum-punch was brought in, and handed round in large glasses pretty freely, and to every glass a toast was given. If at any time we flagged in drinking, "Baron Banks" was always the signal for emptying our glasses, in order that we might have them filled with bumpers, to drink to his health; a task that no Englishman ought to hesitate about complying with most gladly, though assuredly, if any exception might be made to such a rule, it would be in an instance like the present. We were threatened with still another bowl, after we should have drained this; and, accordingly, another actually came, which we were with difficulty allowed to refuse to empty entirely; nor could this be done, but by ordering our people to get the boat ready for our departure, when, having concluded this extraordinary feast by three cups of tea each, we took our leave, and reached Reikevig about ten o'clock; but did not for some time recover from the effects of this most involuntary intemperance. Indeed, we must acknowledge we were somewhat in the same predicament as the guest of the Kamtschatdale, of whom Kracheninnikow farther relates, "Il vomit pendant son repas jusqu' à dix fois: aussi après un festin de cette nature, loin de pouvoir manger pendant deux ou trois jours, il ne sauroit même regarder aucun aliment, sans que le cœur ne lui soulève." On afterwards relating the anecdote of the Stiftsamptman's dinner to Count Tramp, he assured me that he had partaken of a similar one himself, when he first went over to the island, at which time soup was served upon the table made from the boiling down of a whole bullock. Nor are Mr. Phelps and myself the only Englishmen who have suffered from the hospitality of the Geheime Etatsroed; for, since the first edition of this work was printed, I have had the honor of becoming acquainted with Sir John Stanley, at whose table I once had the pleasure of meeting Sir Joseph Banks and Mr. Bright; thus being one of four persons, each of whom, in the course of forty years, had made a separate voyage to Iceland; and each, too, had fresh in his memory the events of the day on which he partook of the feast of the same noble Icelander. I do not recollect the ceremony of the goblet of wine, which, according to Mr. Bright, took place when he and his friends were at Vidöe, but I well remember that the old gentleman made us strike our tumbler-like wine glasses with our finger nail, that we might convince the company, by the vibration of the glass, that we had drunk off the last drop of liquor. At table we were waited upon by two females *, so exceedingly handsomely dressed, that I

* As I had this day, for the first time, an opportunity of observing carefully the dress of an Icelandic lady, which is different from that of other countries, I shall avail myself of the present occasion of describing

concluded they were not common servants, and I afterwards understood that my conjectures were right, and that it was always the custom for the ladies of the house to

it at some length; a thing I am the better able to do, since I had the good fortune to bring one of the richest in the island safe to England with me. I have preserved, also, an Icelandic account of the different articles it is composed of; from an English translation of which, that the governor has been so good as to procure me, I have borrowed a great part of what follows. To begin then with the *Faldur*, or head-dress: this is the most singular and unbecoming part, and I feel such a difficulty in making my description of it intelligible, that I think it right to annex an engraving of it. The inside is composed of a number of pieces of paper, folded into an oblong shape, and neatly covered with two white linen handkerchiefs, in such a way that, below the bottom of the paper, they are formed into a sort of cap, that fits the head, and goes on nearly as far as the ears, which are, however, always exposed, whilst the hair is carefully twisted into a knot on the crown of the head, and entirely concealed. From the top of the head to the extremity, the Faldur measures eighteen inches, and, from a cylindrical shape below, becomes gradually compressed, till the upper part is quite flat, and bends over in the front in a man-

wait at table when any strangers are present. The two who here performed this employment (which is in this country by no means considered a menial one) were, the eldest,

ner that somewhat resembles an ostrich feather, though sadly inferior to it in elegance. Its width at the top is five inches and a half; lower down, near the head, four inches and a half. The part which covers the head is bound round, to keep it on more securely, with two handsome chequered silk handkerchiefs like a turban, but more tight. The upper part is stiffened with numerous rows of pins. Three gilt silver ornaments are fastened to the front of the Faldur, about eight or ten inches above the top of the head, of a spherical shape, hollow, ornamented with open work, and richly embossed; from these hang knobs of the same metal, and rings with leaf-like appendages; in the centre of the ring is an embossed figure of the Blessed Virgin, with our Saviour in her arms. The next article I shall mention is the *Upphlutur*, or bodice; which is made of fine green velvet, bound with a narrow strip of gold lace, with two broad bands of the same materials, and of elegant workmanship, in front, and three on the back; this is fastened before, all the way down the middle, by means of six large clasps of silver gilt, on each side the opening, as large as a half crown, and finely embossed with flowers; and these clasps are rendered more conspicuous by being fixed upon a border of black velvet, with a red edge. From the bodice depends a green petticoat of fine cloth, which goes over several others of wadmal. Over this

the widow of a clergyman, and, the youngest, her daughter, both of whom live in the family, and are maintained by the liberality of our host, who is himself a widower.

is worn another petticoat *(Fat)* of fine blue broadcloth, which, of course, conceals the green one: it is bound with red at the bottom, just above which is a broad border of flowers of various colors, worked in tambour. Over the petticoat in front, is worn an apron *(Svynta)* made of the same materials, ornamented with flowers like the petticoat, and bordered all round with red. From the upper part of it hang three large silver gilt ornaments; the centre one spherical, the lateral ones hemispherical; all hollow, richly ornamented and embossed, and having a silver leaf depending from each, which, together with many of the other ornaments, when the wearer is in motion, contribute no little to making a jingling noise, like horses with bells attached to them. Just beneath these ornaments the petticoat is fastened by means of the *Lyndi,* or girdle, which is nearly five feet in length, and composed of a number of oblong pieces of silver, about an inch and a half long, and one inch wide, sewed with the extremities close together, upon a piece of green velvet, so that it forms a number of joints, and is easily bent round the body, and fastened with a buckle; one end is suffered to hang down in front of the apron, and nearly reaches the bottom of it. All these joints are gilt, and beautifully ornamented with open work, and raised knobs of silver. The jacket *(Treja),* which goes over and conceals a part of the bodice, is made of black velvet,

They were both handsome in their persons, and had beautiful complexions. During the dinner, a large sheep, the finest of the flock, was brought into the room for us to see,

the seams and borders of the sleeves ornamented with fine gold lace, with another stripe of the same down the breast, and gold embroidery near the opening in front, which, at the bottom, is never fastened. but left wide, to exhibit the ornaments of the bodice. The *Kraga* is a stiff and flat collar, an inch and a half wide, completely encircling the neck, and fastened to the upper part of the jacket; this is also embroidered with gold, and sets off the pretty face of an Icelandic girl to great advantage; from the opening in the sleeve hang spherical ornaments, called *Ermaknappa,* of silver gilt, instead of buttons. The *Halstrefell* is merely a piece of white linen put round the neck, over which is bound the *Hals Sikener,* or neck-handkerchief, of purple silk. Around this the *Hals Festi,* neck-chain, three feet and a half long, of silver gilt, and of very curious workmanship, is wound three times, by which means it covers about two inches in depth of the blue silk, and has a very good effect upon it; on one end of it is fastened a large bracelet (*Nisti*) curiously ornamented, and hung round with the initials of the owner: this, also, is of silver gilt. The stockings (*Socka*) of an Icelandic lady are generally of dark blue worsted; the shoes (*Shor*) are made of the skin of seals or sheep: an oblong piece is slit down two or three inches before and behind, and sewed up somewhat in the form of the foot, which it soon takes the shape of by stretching, and is

and was then sent on board our boat as a present. It had horns, was entirely white, and was covered with an extremely coarse and almost straight long wool, intermixed with

drawn tight by a leather thong running along the edge, and tied over the foot. These are so easily made, that I paid only six shillings for a dozen pair. This dress is applicable only to unmarried ladies of rank. To the wedding-dress two rich ornaments are added: one is the *Koffur*, or fillet, worn round the head-dress; it is made in the same manner, and of the same materials, as the girdle, but more elegantly wrought, and the joints are fastened upon gold lace. In the front are the initials of the wearer embossed, surmounted by a crown set with precious stones. The other ornament is the *Herdafesti*, shoulder-chain, made entirely of silver gilt, of considerable weight, and of most exquisite workmanship. This connects seven circular pieces of silver, each as large as a five-shilling piece, and ornamented with silver wire, twisted, and disposed in various figures. The chain is a double one, going over each shoulder, and is terminated behind by a large silver medal, gilt, two inches and three quarters in diameter, and representing in relief, on one side, the crucifixion of our Saviour between the two thieves, with a number of extremely well defined figures below. The superscription is PECCATA. NOSTRA. IPSE. PERTVLIT. IN. CORPORE. SVO. SVPER. LIGNVM. VT. PECCATIS. MORTUI. IVSTICIE. VIVAM̄. The reverse represents Abraham about to offer up his son Isaac, and the angel of the Lord staying the hand already lifted to

shorter hairs.—On returning to Reikevig, Mr. Jorgensen, who had entire possession of the government-house, offered me a bed there, which I gladly accepted.

slay his son: the superscription, PATER. MI. ECCE. IGNIS. ET. LIGNA. VBI. EST. VICTIMA. N. DIXIT. ABRAHAM. DOMINVS. PROVIDEBIT. FILI. MI. I have followed, literally, the spelling of the words: the letters, as well as the whole medal, are in excellent preservation. It is supposed to have been struck in Denmark, and has the date inscribed upon the Altar, 1537. The two ends of the chain are connected in front by a long transverse piece of silver gilt, elegantly embossed and ornamented; from which is suspended a large cross of the same metal, which hangs down upon the breast, and has, in the centre of it, a box for holding perfumes. The lid of this box bears in relief the figure of the Virgin Mary with our Saviour in her arms, and, on the under side, a representation of God the Father, in the likeness of an old man in robes, having a sort of crown upon his head: he is sitting on a throne, and supports with his hands, between his knees, our Saviour upon the cross; while the Holy Spirit, like a dove with outstretched wings, is hovering upon his head: about them are the words VERA. TRINITAS. ET. VNA. DEITAS. Surrounding these, at the four extremities of the cross, are the symbolical representations of the Evangelists. This cross has been in a family in Iceland, upwards of five hundred years. The Koffur and Herdafesti are laid aside after the wedding, and the married lady, in addition to the clothes already described, is

Wednesday, June 28. This was an entire day of rain, so that I rambled no farther than the beach, where a vast quantity of sea-weed was thrown up, principally *Fucus saccharinus*, of which many specimens were six feet long, and one foot wide. Some of the smaller plants had the frond spirally twisted in a very regular manner throughout their whole length; but, on drying them without pressure, the twisted appearance vanished, and they became quite straight.

never to make her appearance abroad without the *Hempa*, an outer coat or habit, of black cloth, with broad borders of velvet of the same color, fastened all the way down before from the chin to the bottom, by means of numerous large clasps of silver gilt, and ornamented with two large circular plates of the same metal on the breasts, richly embossed, and adorned with little leaves, and with the initials of the wearer set in stones. The *Uppslog* are cuffs of black velvet, with gold embroidery. It is needless to say that the Icelandic manufactories do not afford either linen, silk, gold lace, or broad-cloth: these are Danish produce; but all the other articles of the dress are made in the country. Of course, the ornaments of other dresses are not all exactly like what I have here described; but vary according to the fancy of the artist, or the wearer, and few are now to be met with of equal value with these now mentioned.

Thursday, June 29. Another completely rainy day confined me within doors, or to the town. After breakfast a present of butter and Crees' eggs *(Sterna Hirundo)* came from the Stiftsamptman, who at the same time wished to know when I proposed setting off upon an excursion into the country, that he might previously procure me horses and other necessary things. Hitherto, the excessively wet weather had rendered the bogs almost impassable, and the mountains were still every where covered with snow. I therefore determined to wait till this day week before I started. It was proposed that I should go first to the northern quarter of the island, if the weather permitted, and spend some time in Borgafiord, which is reputed the richest and most fertile district in Iceland.

Friday, June 30. Till to-day, the wind had been almost constantly in the south-west, but it has now veered about to the north, and promises a fine and mild day, compared to what we have yet had. With an Icelandic lad for my guide, who went on foot, and frequently faster than I thought it prudent to ride on horseback in such a rocky country, I set out to visit the great bed of *Hraun* (pronounced *Hruin*), or lava, about six miles to the south of Reikevig. The part of it, which I first came up to, was within one or two miles of Havnfiord, where its course has been stopped by the sea, after extending a length of twenty-five miles from the craters, which are supposed to have given birth to this wonderful current. In some parts of the way, there was a track which led us to the spot; but all traces of this track were lost when we came on a small morass, and it was an hour before we reached the Hraun. At a little distance, this huge mass of lava has a most extraordinary appearance, its surface being every where as much broken and as uneven as that of a greatly agitated sea, and its boundaries very distinctly marked by the lighter color of the natural rock, or by the vegetation which this latter produces, whilst the lava itself is almost black, and looks, at a little distance, as bare as if it had issued but the preceding day from the crater. On leaving my horse, and proceeding on foot, with no little difficulty, upon the Hraun, I was still more struck with the strange and desolate appearance that surrounded me. The Etatsroed of Iceland, who was present at the famous eruption of Skaptar-Jökul *, informs me that the torrents of lava, which he had there an opportunity of observing running with a smooth and even surface whilst in a heated and liquid state, in the act of cooling split and broke into innumerable pieces, many of which, of a monstrous size, were, by the expansive force of the air beneath, heaved from their bed, and remained by the side of the chasm which they once filled up. From a similar cause, the whole of this prodigious mass is composed of an infinite number of fragments of melted rock, of various sizes, some twenty and thirty feet high, and of the strangest figures; scattered about an extent of twenty-five miles in length, and of from two and three to ten miles in width, in the

* It ought to have been noticed at p. 6, in speaking of the Icelandic mode of pronouncing the word *Jökul*, that a term very similar is, both as to spelling and pronunciation, applied to mountains of the same kind in Switzerland, where, according to Wagner, they are called *Eis-jöcher*, and that this word *Jöcher*, in Adelung's opinion, is most probably derived from the Latin "*Jugum*".

wildest disorder possible. In appearance, a great part of this lava very much resembles the burnt cinders, or coke, which have been used in drying malt, and is nearly of the same color. The larger masses are generally quite bare of vegetation, but, where the smaller pieces form a tolerably level surface, *Trichostomum canescens* grows in great abundance, and reaches to the length of a foot, or a foot and a half, but is always barren. This, in dry weather, from the numerous colorless hair-like points on the leaves, has almost as white an appearance as snow. Among it I met with the *Geranium sylvaticum*, *Bartsia alpina*, and a few alpine *Salices*, but none in flower. *Fragaria vesca* and *Rubus saxatilis* were coming into blossom. *Encalypta alpina*, which is so rare in our own country, was not uncommon on the lava.

Saturday, July 1. A fine range of mountains to the southward of Reikevig, called the Helgafel mountains, had hitherto been so completely covered with snow, that I knew it was in vain to attempt visiting them. As the snow was now, however, in a measure melted away, and as they did not appear to be at a greater distance than twelve or fourteen miles, I resolved, if possible, to climb some part of them to-day, and accordingly set off on foot, and without a guide, early in the morning for that purpose. But, after going in as direct a line towards the nearest point of them, by the compass, as the nature of the country would permit, at six o'clock in the evening, I found myself, apparently, as far from the object of my walk as when I first set out. This delusion, I apprehended, was owing to the extensive valley that I entered yesterday, through which the lava had made its course, which was concealed by intervening hills from the view of a person looking towards it from the neighborhood of Reikevig. Except for the first three or four miles, the rest of my walk was entirely over the Hraun; and a more toilsome excursion can hardly be conceived: it seemed to be rendered doubly so, by my being obliged to return without reaching the mountains. The immense quantity of *Trichostomum*, which covered a great part of the lava, and filled up the interstices of it, only rendered walking among it more difficult; for it was impossible to see where it concealed a deep hole or a piece of lava, which would give way under my feet; and consequently, I was frequently precipitated upon the sharp edges of the rock. The worst of all was, that I could not well have chosen a more barren spot for plants, in so long a ramble; though I met with one species that delighted me much, and made me for a time forget the fatigue: this was *Andromeda hypnoides* *, which I found just in flower, on the north side of a huge mass of lava, and only there. *Rhodiola rosea* was tolerably plentiful on the Hraun, but scarcely in flower. I also met with *Lycopodium annotinum* and *Conostomum boreale*. In boggy grounds, before I arrived at the Hraun, I found *Orchis hyperborea*, the scent of which is very pleasant,

* Besides the beauty of the color of the flowers of this plant, which particularly attracted the attention of Linnæus, during the course of his travels in Lapland, and induced him to say, that, "florens mirum in modum jucundissimo florum suorum colore spectatorem allicit," it struck me no less forcibly by the singular elegance of its form and general appearance. The delicate tint of the flowers was here finely contrasted with the uniform blackness of the lava. Its barren shoots, as is observed by Linnæus, exactly resemble those of a moss, or of a small *Lycopodium*.

and *Eriophorum alpinum*. On my return, I remarked on the opposite side of a large lake, a small conical hill, of a red color, looking almost as if it were then in a state of fusion. It appeared to me, that, to arrive at this, I had only to go round the east end of the lake, instead of the west, and that, by so doing, I could come into my old track again; but, after walking a great deal out of my way to reach the east end, I met with a deep and rapid torrent, which emptied itself into the lake, and, to my great disappointment, impeded my farther progress. To recompence me, I found growing in this torrent a plant, which I recollected having seen in Mr. Turner's collection, under the name of *Rivularia cylindrica* of Wahlenberg, who gathered it in Lapland, but has not, I believe, yet published it: it grew here seven or eight inches long, and was attached by a small expanded disk to the rocks at the bottom of the stream. Although now not more than half a mile from this little red hill, I was compelled to turn back, and, after getting round the west side of the lake, I hastened to my home, which I reached at twelve o'clock.

Sunday, July 2. This morning Mr. Phelps, Mr. Savigniac, and myself, went in a boat to the Lax Elbe, or Salmon River, a small stream that falls into Reikevig Bay, about six miles east of the town, thus called, on account of the quantity of that fish that frequent it. Mr. Phelps' object was to look at a water-mill, which, he understood, had been erected near the mouth of it; but of which we found little more than the skeleton; for the Danes, who had planned it, never finished the execution of it. During the voyage, we were amused with the number of eider-fowl that were swimming about in all directions, with their young, and we also saw several Swans and Mergansers *(Mergus serrator)*, besides many black Guillemots *(Colymbus Troile)*, and abundance of seals were continually playing within the reach of gun-shot from our- boat. Near the mill, a causeway of stones had been formed across the river, with three openings, in which were boxes for catching the salmon as they return down the river from spawning. Twenty were taken by these means in one night, and so plentiful were they in a pool a little below this spot, that in a few minutes one of our boatmen caught six, by striking a pole, with three barbed points, at them. Three or four others also were caught, by the man leaning over the bank, and suddenly seizing them with his hands. On wet ground, near the mill, *Splachnum ampullaceum* and *Buxbaumia foliosa* were not uncommon. As soon as we reached Reikevig in the evening, we were informed that several persons had called on Mr. Savigniac, to say, that a conspiracy was in agitation amongst the Icelanders, who intended to surround the government-house, and, after having secured such persons as were in it, to take possession of the Margaret and Anne by surprise, as they understood the crew consisted only of twenty-seven men. This tale appeared, at first, too improbable to deserve attention; but, on the arrival of the Etatsroed on purpose to inform us that he had received an offer from fifty Icelanders to join him, if he would raise the same number, and seize upon our vessel, it seemed necessary to take active measures and put a stop to this projected insurrection. Accordingly, Mr. Jorgensen, who had previously placed arms in the hands of eight natives, and formed them into a sort of troop, set off with his soldiers for the house of Assessor Einersen, who was supposed to be one of the chief movers of the conspiracy. A horse was taken for him, upon which he was placed, and, guarded by Jorgensen and his cavalry, was marched, or rather galloped, into the town, and confined for a few days in the government-house.

Monday, July 3. Three days of tolerably fine weather were followed by one of almost continued rain, and, indeed, it was hardly possible to stir abroad the whole week, on account of the wet. I rode, however, one morning, to the hot-spring, where I found a tent pitched, and as many Icelandic women and girls as it could possibly hold, sheltering themselves in it from the weather. They had come with their linen, which was brought on horses from the town, to the hot-spring, where all the clothes of the people, for many miles round, are washed. Some of them had a few little miserable potatoes *, not so large as a

Tuesday, July 4.

* These potatoes, the growth of Iceland, and the best the island afforded this year, were not only wretchedly small, but very bad; not being mealy within, but full of a yellowish tasteless mucilage.

full-sized walnut, which they were cooking in the spring for their dinner, and which they offered me. I had carried with me some eider-ducks' eggs, for the purpose of trying how long it would take to boil them hard, and I found they required ten minutes, whilst lying in a part of the water where the thermometer rose to 200°.

Saturday, July 8. After a stormy night of wind and rain, the weather cleared up about nine o'clock, and, being furnished with horses, tents, &c., and a guide, by the Stiftsamptman, I set out for the Geysers, which I proposed visiting before I went into Borgafiord. This I was the more anxious to do, as it seemed probable, from the many unlucky events which happened, and were inimical to the trading between the Icelanders and the English, that we should not make any long stay, and Mr. Phelps was very particular in desiring me to come back at the expiration of a fortnight at latest, lest the vessel should be ready for sea; for that there was no prospect of my getting to England this year, if I did not return with the Margaret and Anne; since the Flora, a ship of Mr. Phelps', whose arrival he expected soon after our own, was not yet come, and no other British vessel was expected. Three horses were loaded with tents, provisions, &c., and a fourth was a relay. These were fastened to each other in a line, by a rope of twisted horse-hair, tied at one end to the tail of the first horse, and, at the other, to the under jaw of that which was next to it; and so on with the rest. My guide rode before, holding a line, fastened to the mouth of the first luggage-horse, so that they all followed exactly the same track, and, so accustomed are these horses to this mode of travelling, that, even when they are not tied, they will still keep following each other, to the great annoyance of any person who may happen to be riding them, and may wish to go a little faster than the rest, or to leave the regular line. A man from the ship, of the name of Jacob, who, although a German by birth, understood sufficient of Danish to act as interpreter between me and an Icelander, who spoke that language, rode a sixth horse, and I a seventh; yet, even these, numerous as they may appear for one person, were found not sufficient for our journey. There is, for some distance from Reikevig, a sort of beaten way, along which we went with greater ease than I had expected. Before we arrived at the doors of the first house we met with, the inhabitants came out to offer us sour whey and milk, in large wooden bowls, carved with no other instrument than a knife from birch-wood, and covered with a lid, on which, and sometimes on the two ears, are cut leaves and other ornaments. They hold about a pint and a half, and are used by all the natives to carry their butter, when they go upon a journey, as well as to drink their whey and milk out of, when at home. These good people were examining, with great attention, a pistol, which Jacob had slung at his girdle, and which they were very anxious to know the use of; but this it was not so easy to explain to them, nor would it have been prudent in our present situation, when we might be called upon to make use of it in our own defence, against these very persons, whose ignorance was our surest protection. From this place, which I understood was named Kirkat, and which lay due east from Reikevig, we took nearly a northerly course, in our way to the head of

Thingevalle-vatn, or the Lake of Thingevalle. The weather was so rainy and thick, that we scarcely saw any thing of the country, till we arrived at the base of the mountain, Skoul-a-fiel, whose three lofty and cone-shaped summits are plainly seen from Reikevig, and by far exceed in height any of the neighboring hills. At the foot of this mountain, a deep and narrow chasm caught our attention, which seemed as if it had been formed by some violent convulsion of nature, and continued for some way by the side of our road. Near it, I also remarked the perpendicular side of a hill, composed of basaltic columns, jointed here and there, like those in Staffa, but not more than eight on ten inches in diameter, and less regularly columnar. From this place, till we got to the banks of the Lake of Thingevalle, nothing interesting occurred. The country, through which we passed, consisted either of a dreary moor, over which large masses of rock were every where scattered, or of a disagreeable morass, into which our horses not unfrequently sunk up to their bellies. In one of these morasses, I passed a woman, driving a horse, loaded with the trunk of a tree, which had been dug up close by: it was so large as to appear nearly as great a burthen as the beast could well walk under, and was, probably, five or six feet long, and nearly a foot in diameter. I do not recollect meeting with any remarkable plants, different from those I had before seen about Reikevig, except an *Orchis*, with a singularly inflated and semi-transparent nectarium, of which I could find no description in the *Flora Scandinaviæ*. Several sorts of dwarf willows were common, as well as *Bartsia alpina*, *Geranium sylvaticum*, and *Conostomum boreale*. When we reached about half way of our day's journey, we stopped half an hour to bait our horses, and arrived at Heiderbag, where we proposed to remain the night, between ten and eleven o'clock. The priest Egclosen, at whose house I called to deliver a letter from the Stiftsamptman, rose from bed, and assisted us to fix our tents, and unload the horses; but the heavy rain had wetted almost every thing, so that we passed but an uncomfortable night, lying in our damp clothes, and on the moist and swampy ground, where our tents were pitched.

Sunday, July 9. Early this morning, the priest came to invite us to breakfast at his house, which I readily agreed to, taking with me tea, coffee, and other provisions; a precaution absolutely necessary, for his house would afford nothing but milk, skiur, butter, and fish. I was even obliged to send back to my tent for a kettle to boil the coffee in. The only part of the house to which we were admitted was that in which the fish, tallow, wool, milk, &c., were kept; for this, being the best part of an Icelandic building, is used for the reception of strangers. It had walls of alternate layers of turf and stone, without either cement to unite them, or plaister to conceal their nakedness, and the floor was the bare earth. One chair was all our host could furnish, and, indeed, there would not have been room for more, so completely was the place lumbered up with old chests, old clothes, &c. What little provision there was in the house was most willingly offered, and it was with difficulty I could prevent him from killing a lamb, to entertain us better. This man had been secretary to the Stiftsamptman, who had procured for him the curacy of Thingevalle (there being no church at Heiderbag), which would be the means of his ultimately obtaining a more lucrative situation. At present, his income is extremely narrow, being only six rix-dollars a quarter (twenty-four shillings) from government, but the marriage and burial fees amount to something more; the former ceremony, I think, is performed for two marks: in addition to that, he has a house to live in free of expence, and some glebe, which enables him to keep five cows, and twenty-eight sheep. Three miserable cottages, also, stand upon his glebe, for one of which he receives four dollars, for another three, and for a third two dollars per annum. The chief employment of the female part of his family, besides knitting, is making butter, skiur, and sour whey, which constitute almost their only food. In the winter, if the weather is very severe, the priest is obliged to kill some of his cows and sheep, for want of a sufficient quantity of hay, and in such cases, only, can they afford to live upon flesh. After breakfast the priest visited his nets in the lake, which had been se for the first time for catching a fish, which the Danes call *Forelles,* and which is allied to our *Char*, but, I think, quite distinct. Although I compared it accurately with the descriptions of the various species of *Salmo* in *Shaw's Zoology*, which I had with me, I could not find that it agreed with any of them. Only one was caught, which we cooked, and found very delicious *. At noon our friend was obliged to take leave of us, as he was under the necessity of setting off for Reikevig, where he was to preach a sermon before the bishop on the following (Monday) morning. He assured us, however, as there was every appearance of a continuance of the rain, which fell in torrents the whole day, and of our being consequently detained, that he would, if possible, be home the following day, that he might accompany us to Thingevalle, where his principal, as he called him, lived, and would receive us kindly. We hardly expected to see him return at the time ap-

* The season of the year in which the *Forelles* abound in Lake Thingevalle was now approaching: about the 29th of July they are caught in the greatest plenty, and of a large size, some of them weighing from ten to fifteen pounds.

pointed; for, in addition to his own weight, his horse had to carry two large chests, containing tallow, wool, and worsted stockings, which were to be bartered for iron and other articles of necessity, at Reikevig.

Monday, July 10. A little better weather this morning induced us to put our luggage out of the tents to dry; but this was scarcely done when it began to rain, and continued to do so, without intermission, the whole day. We were not even able to light a fire, but were obliged to send our provisions to the priest's house, which was full a quarter of a mile off, to be cooked.

Tuesday, July 11. After a night of wind and heavy rain, about ten o'clock the weather cleared up, and, with the exception of a few showers, was fine during the remainder of the day. A brighter atmosphere now permitted us to catch a glimpse of the neighboring scenery; and the first thing that drew our attention was the immense Lake of Thingevalle just before us, and seeming as if placed there by enchantment, as, though almost at our feet, we had hitherto seen nothing of it, except the margin. It is reckoned fifteen miles long, and from five to twelve miles wide. Near the middle rise two fine black insulated rocks, of considerable size and height; the largest called Sandey, and the smaller one Nesey, upon both which, thousands of the Black-backed Gulls *(Larus marinus L. Svart Bakr Isl.)* annually rear their young. North and south of this lake, were some grand rugged mountains, but at a considerable distance from the place in which we were, and mostly covered with snow. Whilst we were looking at this magnificently wild scenery, the priest came down to us, having returned late the night before, after a journey of two days on horseback in incessant rain, during which time he did not once change his clothes; not even when he had to preach before the bishop. We now proposed taking a walk by the side of the lake, and setting off on our journey early in the afternoon. The margin we found every where flat, and the water appeared extremely shallow for a considerable way into the lake, but it is by no means so towards the middle, where, in some places, the natives cannot fathom the depth. The shores and the bottom, as far into the lake as we could see, were formed of small black fragments of rock, except that in a few places, at a little distance from the edge, there are some entire and romantic masses, on which I found several mosses that I had not before met with in Iceland: some of them, indeed, were quite new to me. A beautiful *Lecidea*, with a white and powdery crust, and red shields with an elevated margin, grew in small patches upon so hard a substance, that I was not able to procure the smallest piece. In the lake was abundance of *Rivularia cylindrica*. At four o'clock we set out, accompanied by the son of the priest of Thingevalle and by the priest Egclosen, for Thingevalle, which was only at the opposite side of the head of the lake, and not more than five or six miles distant; yet, owing to the badness of the road, and to our stopping to look about us, it was eight o'clock before we reached it. Nearly our whole ride lay along the shores of the lake, which are composed entirely of small broken pieces of lava, in many places nearly as fine as sand, and as fatiguing to the horses as sand itself would have been. Among this, wherever the numerous streamlets, which ran into the lake, had deposited a small quantity of soil, the bright yellow green of *Bartramia fontana*, and the pink-colored flowers of *Sedum villosum*, were finely contrasted with the blackness of the ground. In some places, at a short distance from the shore, such of the rock as had been melted was in an entire state, and marked on the surface all over with numerous elevated semicular lines, in a manner not unlike the shell of an oyster *, if such a comparison may be allowed. We passed a tolerably wide stream, just below a cascade of considerable size, which reminded me of the upper fall of the Clyde; but there were no trees, and scarcely a blade of grass, to clothe the surrounding rocks. Having reached the north-eastern extremity of the lake, our guide told us we were

* As a figure will give a better idea of this appearance than words can possibly do, I will beg to refer, for an excellent representation of this kind of unbroken lava, to plate 35 of *Bory de St. Vincent; Voyage dans les quatre principales Isles des mers d'Afrique.*

coming to the pass of Almannegiaa, which I had heard much of, as one of the greatest curiosities in Iceland. We already found the ground broken into a number of great openings, of various length and width; some so deep, that the darkness prevented our seeing the bottom, which in others was concealed by ice and snow. On a sudden we came to the brink of a frightful precipice, down which we looked into Almannegiaa, a monstrous chasm, extending almost as far as we could see, in a direct line, nearly east and west: through this our road lay. A smaller opening branches off in a south-east direction, and, a great number of large pieces of rock having fallen into it, the natives, without any assistance from art, make it serve as an entrance to the other. Here, however, we were obliged to have all the luggage, even the saddles, taken off our steeds, and carried on the shoulders of our people. The horses were then driven down between the great stones which composed the descent. A more rugged pass * can hardly be con-

* "Ce chemin est aussi dangereux que difficile; il y a une infinité de degrés *taillés* dans le roc, par où les

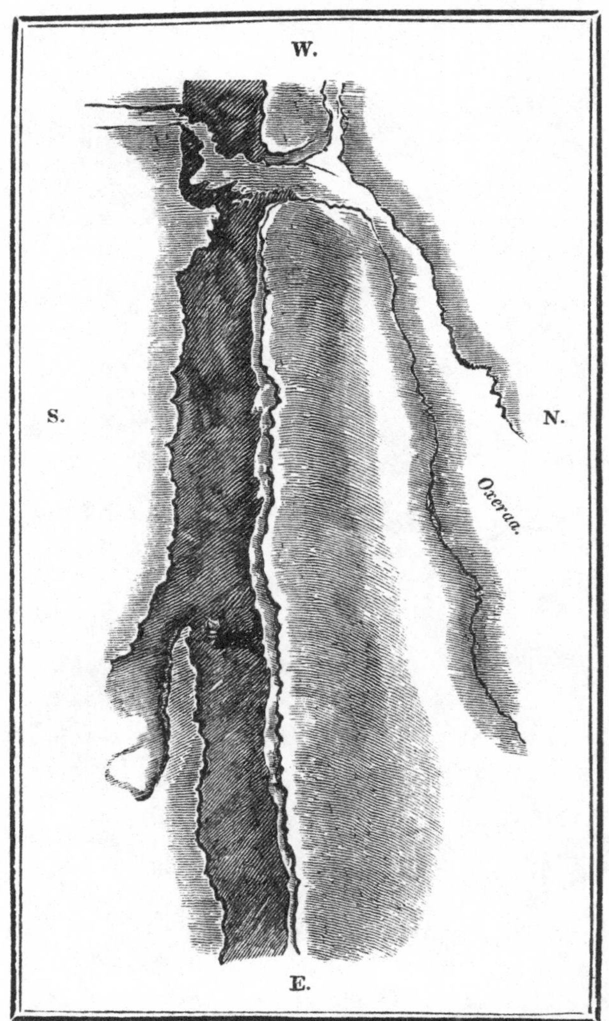
W.
S.
N.
Oxeraa.
E.

ceived. As we descended by this rude but natural staircase, the sides, which were perpendicular, became proportionably higher, till, winding round some huge fallen pieces of rock, we entered the great chasm. A grassy bottom of considerable width, and extending as far as we could see, afforded a sufficient, though not a very luxuriant, pasture for our horses; and this determined me to have our tents fixed here, that we might remain all night in this remarkable spot, some idea of the ichnography of which I have endeavored to convey by means of the annexed engraving, which, however, represents it so imperfectly that I omitted it in the first edition of this Tour, and am fearful my readers may think I might as well have done so in the present. On the left of the entrance to my tent, rose a perpendicular

hommes grimpent, et mènent leurs chevaux, qui montent ces degrés, en faisant des sauts qui ne les avancent pas toujours."—*Povelsen and Olafsen,* § 863.—I presume, by the word *taillés,* Messrs. Povelsen and Olafsen do not mean to imply *cut by art;* for I certainly could not perceive that any artificial means had been employed, nor could they have been so to advantage, without more powerful engines than the Icelanders are possessed of.

wall, above an hundred feet in height, blaek and craggy, with here and there a little vegetation, and a stunted birch, which took root among the ledges of the rock: it was on the lofty summit of this that our priest told us criminals used to be executed *: on the opposite side, and at about the distance of twenty yards, rose another wall, equally perpendicular, and more craggy, but not half the height of the former, yet, probably, in consequence of its being less exposed to the rays of the sun, covered with a more abundant vegetation, especially of moss *(Trichostomum canescens)* and *Saxifrages*: about a hundred yards from us in front, a little bend, in the direction of the chasm, appeared to shut us in by a lofty precipice: behind us was the pass or entrance to the chasm, which I have just described, and by the side of it a continuation to the southward of the high walls of the chasm; but the passage was almost choaked up by a

* On looking into the French edition of *Povelsen* and *Olafsen's Travels,* I find the above place mentioned as "la roche escarpée d'où l'on précipitait jadis, dans le bûcher, les victimes condamnées à être brûlées pour crime de sorcelerie." *Tom.* v. *p.* 363.

vast number of loose pieces of rock, which had fallen from the precipices above. However, we had now no time to examine the place more; for it was necessary to pay our respects to the priest of Thingevalle, who lived scarcely a mile from the place. We therefore left our luggage and tents in charge of the guides, and, going northward in the chasm, came to a little opening on the east side, through which we had to pass. Having reached this, we looked down into an immense plain, which was every where intersected by rents in the earth, as far as the eye could reach, crossing each other in various directions, though most of them were torn from north to south: three in particular seemed to extend, in uninterrupted lines, the whole width of the plain, and were terminated on one side by the lake Thingevalle. Immediately below us was the river Oxeraa, and, just on the other side, in the midst of this most extraordinary country, are situated the church and parsonage of Thingevalle *. The verdure upon these buildings, and the unusual

* This place takes its name from the word *Althing*, or the seat of the court of justice, which was once there, but was before that time, according to Povelsen and

fertility of the small patch of ground which immediately surrounded them, together with the numerous herds of cattle, made a pleasing contrast with the rest of the country, which was, as the French editors of Povelsen and Olafsen term it, "horriblement bouleverseé par le feu souterrain." We went out at the above-mentioned opening, and, crossing the Oxeraa, arrived at the parsonage by a road fenced in on each side by a low

Olafsen, at Kialarnoes, and is now at Reikevig. The Oxeraa divided the *Althing* into two parts: the consistory, which was upon the eastern bank, was held every year in the church of Thingevalle, but only for the bishoprick of Skalholt; for the northern bishoprick, the consistory was held at Kugemire, in the canton of Skagafiordur. Upon the western bank of the river was situated the building, made use of for the session of the inferior court, called *Lavretten*. The *Lavretten* was held in the open air till 1690, when a building was constructed similar to the rest, belonging to the *Althing*, that is to say, with walls of lava, and a roof covered with rafters and laths, ornamented on the outside with wadmal. Thorleosholm, a little island in the river Oxeraa, was the place of punishment for the criminals.—See *Povelsen and Olafsen*, § 905.—Tingwall is, also, the name of a place in the Shetland Islands, where formerly the chief court of justice was held.—See *Mr. Neill's* interesting *Account of the Orkney and Shetland Isles*, and *Edmonstone's Zetland Islands*.

stone wall. A fine pair of rein-deer's horns, fastened against the side of a building here, particularly caught my attention. These animals were first introduced into this country (according to Von Troil) in the year 1770, from Norway, by order of Governor Thodal. Of thirteen then sent ten died on the passage. The three remaining ones have done extremely well, and bred so fast, that at this time Count Tramp reckons that there are about five thousand head in the island. They are, however, quite useless to the natives; for no attempts have been made to domesticate them, nor can the inhabitants afford to buy powder and ball to enable them to kill them for provision. They herd together in the wildest and least frequented parts of the mountains, where they are seldom seen, and are not shot without extreme difficulty. It seems truly extraordinary that, in a country so wretchedly poor as Iceland, and so ill calculated for the subsistence of the greater number of useful quadrupeds, the rein-deer, which is peculiarly adapted to their Lichen-covered plains, should be allowed to wander at large, not only unserviceable to the natives, but devouring a plant which serves themselves in part for nourishment, and is also of importance as an article of export. This too, when the Laplanders, seated nearly in a similar country and under the same latitude, find in these animals the blessing of their lives. Could they but be persuaded to see and to follow their true interest in this respect, to them might be applied what has been so beautifully said of their neighbors, * "Hi Lichene obsiti campi, quos terram damnatam diceret peregrinus, hi sunt *Islandorum* agri, hæc prata eorum felicissima, adeo ut felicem se prædicet possessor provinciæ talis sterilissimæ atque Lichene obsitæ. Pecora enim bene perferunt clima illud; habent sufficiens alimentum; reddunt pastori et vestimenta et alimenta." We found the priest, who was the object of our visit, smoking his pipe † in the front of his house, surrounded by his wife and numerous domestics, who had all come out to gaze at us.

* *Fl. Lapp.* p. 347.

† This is a luxury in which only the richer Icelanders can afford to indulge. A pipe in the mouth of an Icelander is, therefore, not a common sight, and is mostly confined to Reikevig, where they learn the custom from the Danes, who are always smoking.

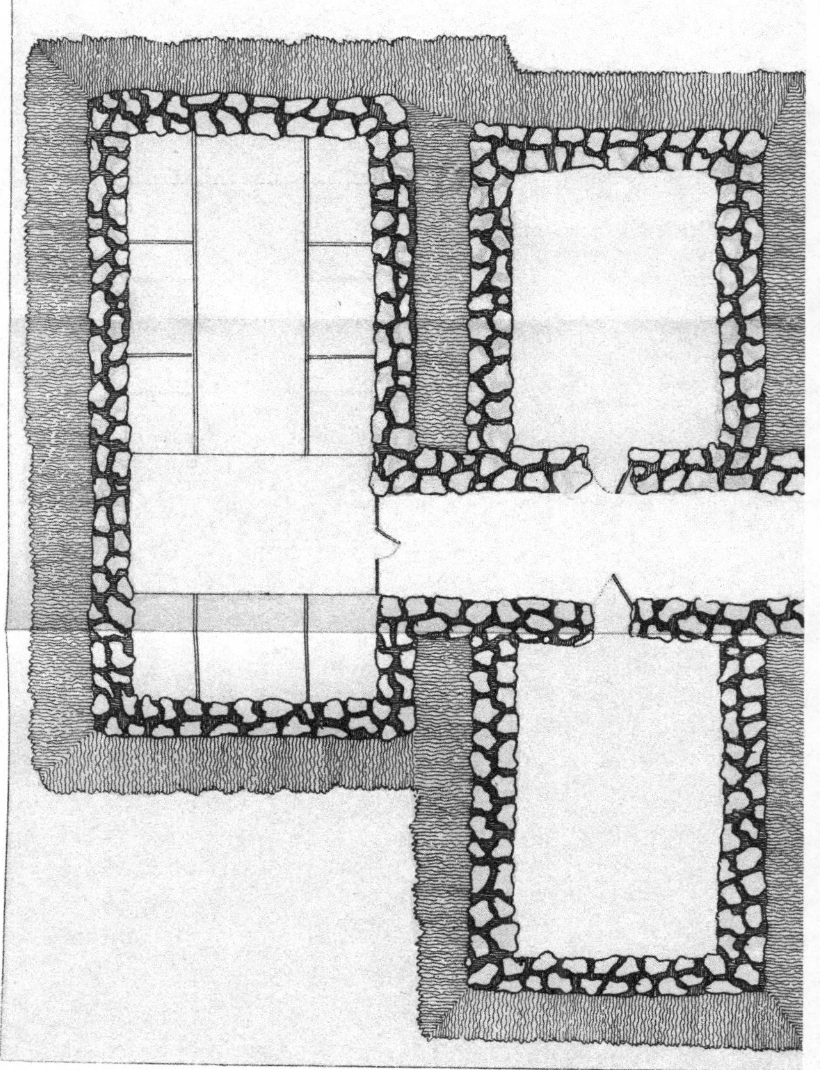

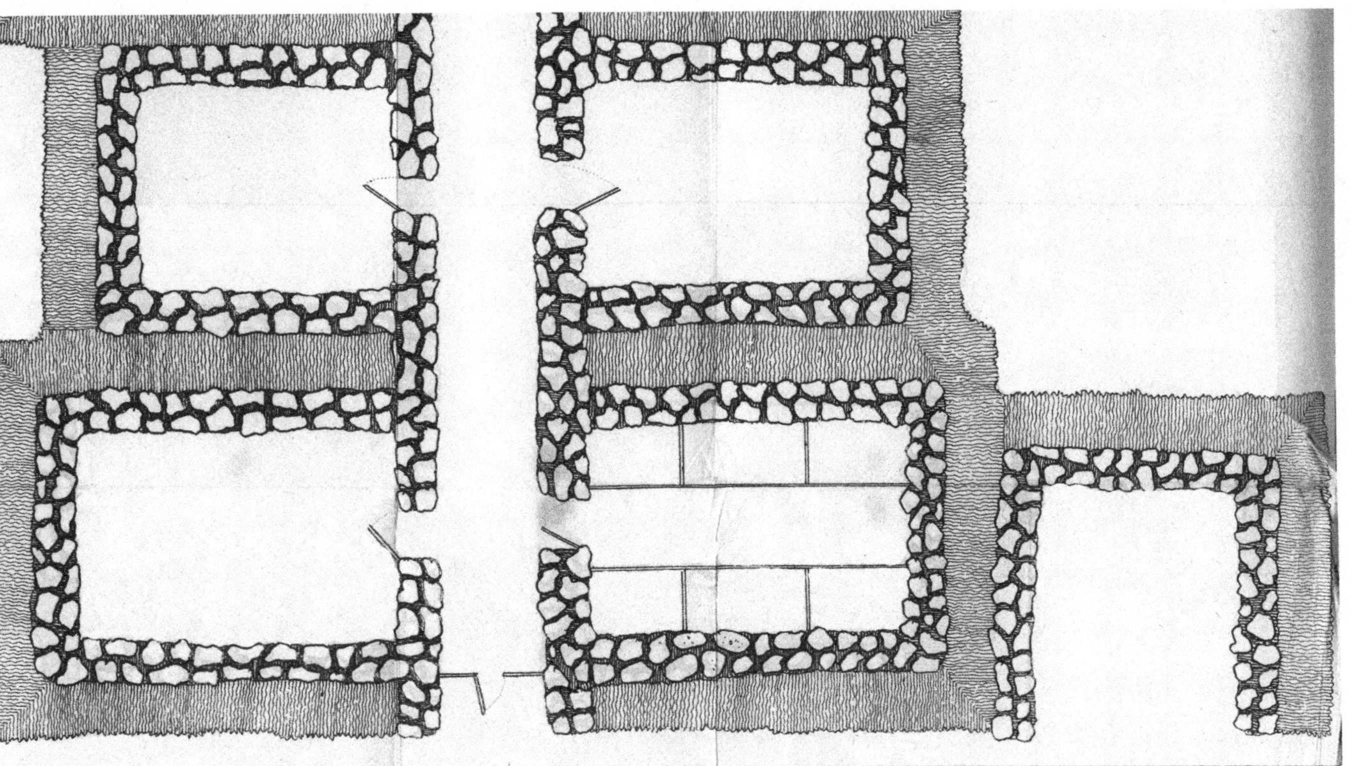

Plan of an Icelandic House

His dress bespoke but little of the clergyman, not differing, that I could perceive, in any respect from that of an Icelandic peasant. He even wore the common blue cap, which concealed but a small portion of his white and venerable hair that hung over his shoulders. He offered us milk, fish, or any thing that his dwelling afforded, which could be of service to us. His residence was a pretty good one, and more extensive than is common in Iceland, where, generally, a low fence of stone or turf encloses a considerable portion of ground, and, in the midst, stands a cluster of little buildings or cabins, which, taken collectively, constitute an Icelandic house: the walls, formed of alternate layers of stone and turf, are extremely thick, especially at the base, and do not stand perpendicularly, but lean a little inwards: their height is about seven or eight feet; and the addition of a sloping roof of turf, laid on birch boughs, raises the whole edifice to twelve or fourteen feet. It is to be observed, that to all these, except one building (which is, nevertheless, united by walls to the rest), a single entrance serves: so that, going along a strait passage, as narrow as it is damp and dark, you come to others which branch off to the right and left, and communicate to the different chambers or rather cabins, of which the whole house * is composed. One or two are occupied as sleeping-rooms, where two or more beds, elevated about four feet from the ground, are placed by the side of the wall, the head of one touching the foot

* Sir George Mackenzie gives the following strongly-drawn picture of an Icelandic house, which, unfortunately, is applicable to too many of them: "the thick turf walls, the earthen floors, kept continually damp and filthy, and the personal uncleanliness of the inhabitants, all unite in causing a smell insupportable to a stranger. No article of furniture seems to have been cleaned since the day it was first used; and all is in disorder. The beds look like receptacles for dirty rags, and when wooden dishes, spinning-wheels, and other articles are not seen upon them, these are confusedly piled up at one end of the room. There is no mode of ventilating any part of the house; and as twenty people sometimes eat and sleep in the same apartment, very pungent vapors are added, in no small quantity, to the plentiful effluvia proceeding from fish, bags of oil, skins, &c. A farm-house looks more like a village than a single habitation. Sometimes several families live enclosed within the same mass of turf. The cottages of the lowest order of people are wretched hovels: so very wretched that it is wonderful how any thing in the human form can breathe in them." *Travels in Iceland*, p. 115.

of another. The bedstead is made of boards, and has high boards on the side, so that, except in being larger, it differs but little from such as are frequently seen in ships' cabins. Curtains, and all other kinds of bed-furniture, are unknown. The beds themselves are either of down, or are merely a loose heap of *Zostera marina*, over which are thrown three or four thick coarse pieces of wadmal. One room is appropriated to the loom, another serves as a sitting-room, and a third as a kitchen, where the fire is made of turf, or, as is the case at Thingevalle, of small twigs of birch. Sometimes, also, the same entrance leads to the dairy, but the priest of Thingevalle had his in a detached building, differing, however, in no respect from the rest, where the milk and cream were kept in large square shallow wooden troughs, standing upon stools all round the apartment. The fish-house, in which, besides the dried fish, wool, clothes, tallow, saddles, and the few implements of husbandry are placed, is considerably larger than the other rooms, to which, however, it is united, but has a separate entrance. The fronts of all these places resemble the gable ends of English houses, and are formed of unpainted boards, standing vertically. With regard to the interior, both the sides and bottom are but seldom boarded: the former are usually nothing but the black stone and turf, and the latter only the bare ground. Generally, there are small openings, either in the walls or roof, by way of windows; but these are rarely glazed, and more frequently covered with the amnion of the sheep, which allows but a small portion of light; yet even this is a luxury, and is to be found only in one or two of the rooms. A chimney, or rather an aperture for the emission of the smoke, usually made with a tub, is seen in the best houses alone: in others the smoke is left to find its way out at the door, by which, also, the only air that they can possibly receive is admitted. The son of the old priest accompanied us in a walk among the neighboring chasms; which are, every where, so numerous, that we could scarcely go ten feet without coming to the edge of one that barred our farther progress in that direction. Some at the bottom have snow and ice, others contain the purest water that can possibly be conceived, so deep, that in many places no bottom is to be found, and at the same time so clear, that, on throwing in a stone, its descent may be traced with the eye for a considerable length of time. We saw abundance of small fish swimming here, some of which we caught, and found to be the young of the Thingevalle trout; so that, although at a considerable distance from the lake, in all probability some of the subterraneous caves which abound, together with the chasms, all over this district communicate with it. A little herbage covers the intermediate spaces between the clefts, but the more common alpine Lichens and Mosses occupy the greater part of the surface. *Dicranum purpureum* astonished me by its size and abundance. In some of the caverns, among the drippings of the rock, several plants of *Veronica fruticulosa* were displaying their lovely blossoms, and, on the edge both of the caves and precipices, *Polygonum viviparum* grew in such profusion as to form thick tufts, several feet in diameter, and of great size. Cattle are often sent here to graze, but not without the annual loss of several, which fall into the holes and perish. The priest Egclosen had himself a narrow escape from death, having one evening slipped into a chasm that was half filled with snow, where he remained till the next morning, when he was searched for, and, fortunately, discovered in time to save his life. On returning to the house, we found the women and girls milking the sheep, which were for this purpose enclosed in a large oblong four-sided wall, made of lava and turf, in alternate layers, with a door for the admittance of the women, and a small square opening, just high enough to permit the sheep and lambs to be driven in: a still smaller one communicates with another little enclosure, into which, through this aperture, which is not large enough to admit the sheep, the lambs are put, whilst the mothers are milked; otherwise, they would be restless and unwilling to stand still. Many of these sheep afforded a quart of milk, of a rich quality, but that which comes of the second milking, is, by far, the best; for it is the custom here, having milked the whole flock, to begin again and milk them a second time. The cows are milked in the open ground, with their hind legs tied by means of a horse-hair line. From the milking-place, we visited the church, which stands upon a little eminence, at a short distance from the minister's dwelling. It was of a simple construction; in form, an oblong quadrangle, with thick walls, leaning a little inwards, composed also of alternate layers of lava and turf. The roof was of turf, thickly covered with grass, and, from the top of this to the ground, the building was scarcely more than sixteen or eighteen feet high. The entrance end alone was of unpainted fir planks, placed vertically, with a small door of the same materials. I was surprised to find the body of the church crowded with large old wooden chests, instead of seats; but I soon understood that these not only answered the purpose of benches, but also contained the clothes of many of the congregation, who, as there was no lock on the door, had at all times free access to their wardrobes. The walls had no covering whatever, nor had the floor any pavement, except a few ill-shapen pieces of rock, which were either placed there intentionally, or, as seems most probable, had not been removed from their natural bed at the time of the building of the church. There was no regular ceiling: only a few loose planks, laid upon some beams, which crossed the church at about the height of a man, held some old bibles, some chests, and the coffin of the minister, which he had made himself, and which, to judge from his aged look, he probably soon expected to occupy. The whole length of the church was not above thirty feet, and about six or eight of this was parted off by a kind of screen of open work (against which the pulpit was placed) for the purpose of containing the altar, a rude sort of table, on which were two brass candlesticks, and, over it, two extremely small glass windows, the only places that admitted light, except the door-way. Two large bells hung on the right-hand side of the church, at an equal height with the beams. I observed that the Icelanders pull off their hats, on entering their place of worship. We left our friend Egclosen to take his rest at Thingevalle; but, preferring to sleep in the tent myself, after being abundantly supplied with trout and milk, we returned to Almannegiaa. On walking to the north of the chasm, I met with a few scarce plants: among them were *Carex atrata*, extremely fine, *Saxifraga rivularis*, *Veronica fruticulosa*, *Osmunda lunaria*, *Polypodium arvonicum*, and *Hypnum silesianum*. I much regretted not being able to spend more time here; but, as a visit to the Geysers was the principal object of my journey, I thought it best to accomplish that first, and, if there were leisure, to wait here a few days on my return: we therefore proposed, should the weather be suitable, to continue our route early in the morning.

Wednesday, July 12. The morning proved fine, and we had scarcely breakfasted, when Egclosen and Thorlavsen (son to the priest of Thingevalle) called us to proceed on our journey. They both kindly offered to accompany me some way, that they might point out such objects as were most worth our attention. We stopped at Thingevalle, to take leave of the priest, and, having refreshed ourselves with some rich cream which he offered us, we then pursued our course in a south-easterly direction, among the innumerable cracks, rents, and hills of rugged lava, which rendered travelling extremely fatiguing for the horses, and by no means free from danger; for a false step, or a rolling stone, would infallibly have precipitated both the animal and his rider to the bottom of a chasm. The passages between many of these openings were scarcely of sufficient width for a single horse, and were, also, so full of holes, that it required beasts used to this country to attempt to go along them; but the most fatiguing part of this day's journey was when we had to traverse the three long chasms, which I have already mentioned as extending across the plain. They were of considerable depth every where, except in the parts where we crossed them, and, there, they were half filled up with loose pieces of lava, forming a rude natural causeway. At the entrance of one of these *, we were again obliged to have all the luggage taken off the horses, and carried over on mens' shoulders. We were then full half an hour in crossing a place of not more than two or three hundred yards; except that we were occupied some little time, in helping the horse of the priest Egclosen from a hole, into which he had

* Called Hrafnagiaa. Povelsen and Olafsen, speaking of the numerous openings in the ground about Thingevalle, say, "Celle de Hrafnagiaa embarrasse sur-tout beaucoup les voyageurs; parce qu'il y a bien peu d'endroits où l'on puisse la passer ou la traverser."

fallen among the rocks, and where he had torn the skin more than half way down his leg. This misfortune, which lamed the poor animal considerably, and which, to a native of any other country, who, like this man, was worth only one horse in the world, would have been a cause of uneasiness, if not of complaint, had no such effect on Eglosen: he did not repine at what had happened, but went cheerfully on his way, with his limping and bleeding horse, only observing on the accident, that "it could not be helped, the place was so bad." I know not whether it arises from a peculiar resignation to the will and providence of God, produced by real piety, or whether it is ascribable to the effect of climate, and to the poverty and distress which attend upon the whole life of the Icelanders, that they seem to feel less for the calamities of themselves or of whatever surrounds them, than is the case with any other people I have read of. When I was lamenting the number of lives, which, Egclosen informed me, were lost among the holes that are here every where met with, he stopped me by saying, "it is God's will that it should be so." On arriving at the opposite side of the chasm, we found ourselves in a somewhat better track, but, as our friends from Heiderbag and Thingevalle were not thoroughly acquainted with this country, it was recommended to us to call at a peasant's house, which was but little out of the way, where we might procure something to apply to the leg of the wounded horse, and at the same time might inquire after a guide, who would be able to direct us to some remarkable caves in the neighborhood. We were disappointed, on reaching the cottage, to find there was only an old woman at home, who, nevertheless, made us welcome, and immediately produced some excellent milk for our refreshment, and some *syre,* or sour whey, which answered both for washing the horse's wounds, and for drink to our guides. In the absence of the male part of the family, the woman undertook to be our conductor, and, without either shoes or stockings on her legs and feet, with extraordinary agility, sprung cross-legs upon a spare horse that we had, though destitute of saddle and bridle, and took the lead of our little cavalcade. She pointed out to us the entrances to several large caves, one of which in particular, called Undergrandur, is said to penetrate a considerable way into the ground. We alighted from our horses, and went in as far as we thought it prudent without lights. The entrance was about ten or twelve feet high, and about twice that width, but both the height and width increased as we advanced. For some way in, the snow had been drifted, and still lay unmelted, intermixed with ice. Beyond this, vast black pieces of rock, of an enormous size, covered the bottom, and similar ones hung suspended from the roof, which seemed to threaten every minute to add to the number of those below. We climbed over the heap upon the ground, and groped our way, till we almost lost sight of the light at the entrance. Darkness prevented our proceeding farther, and the coldness of the place, and dampness owing to the constant dripping from the roof, made us glad to return to the open air. We looked into two or three other caves, but attempted nothing more; as their appearance presented nothing particularly interesting, or likely to repay the trouble and hazard of investigation, they being mostly barren of all vegetation, and dark. At the mouth of one I found a miserable specimen of *Andromeda hypnoides*, and a few plants of *Pyrola minor*. Our female guide now took leave of us, after having given us directions for our route, which lay almost entirely among broken lava. We had not proceeded far, when Egclosen told us that we were drawing near the crater of a volcano, and recommended to us to leave our horses, as it would not be easy to approach it with them, and walk to the spot. Following this suggestion, we quitted a somewhat level tract of fragments of lava, heaped one upon the other, and came on a gently rising eminence of no great elevation, but composed of a more solid mass, cracked, indeed, into innumerable pieces, but these were still lying in their original bed, and not at all scattered about: the surface was tolerably smooth, except that it was marked with elevated semicircular lines. The summit of this hillock was terminated by a still more solid mass of rock, of nearly a conical shape, all consisting of calcined matter, which had evidently been formed from the melted rejectamenta of a volcano; indeed, this was the rim or mouth of one, and elevated about ten or twelve feet from the above-mentioned lava. On climbing to its top, we found the edge extremely rugged, sharp, and vitrified, having an orifice from six to seven feet wide, and gradually becoming narrower for a few feet as it descended, then widening again, and forming a hole, whose depth I was by no means able to ascertain. That it did not descend exactly in a vertical direction for any great length of way, was made evident by throwing in a stone, which soon struck upon some projecting ledge or bend in the pipe. The color of this cone on the outside was a deep greyish brown, almost inclining to black, and in some places a full red, considerably darker than the lava it stood upon, which appeared to have been exposed to a less degree of heat. There was no smoke, nor any smell of sulphur to be perceived; nor, to judge from the grass that grew in thick tufts some way down the crater, had there been any for a great length of time. The natives, too, had no tradition of its having thrown out fire, neither was the place itself known to many who lived in this quarter of the island. Sir John Stanley seems to have passed over a part of this same bed of lava, during his travels, and was at a loss to imagine whence such a prodigious mass could have issued. I should have been equally so, if it had not been for the friendly priest Egclosen, who alone, of several Icelanders now with us, was acquainted with this crater, which undoubtedly gave birth to a portion, at least, of the lava that surrounds it. Having spent some time here, and made a few sketches of the spot, as well as the violence of the wind would allow me, we took leave of Egclosen and Thorlavsen, and continued our journey. We descended from the little eminence on which the crater stood, and arrived in a short time at the foot of a great mountain, whose sides appeared entirely composed of fragments of bare rock, varied, indeed, between the interstices with patches of *Trichostomum;* but these of small size, and scattered at not small intervals: near the summit the snow lay in considerable quantity, over, perhaps, a solid bed of rock *. As we passed round the foot

* I have observed mountains in Iceland more lofty than this, composed entirely of loose pieces of rock, with their summits perfectly free from snow; whilst others in their vicinity, of much less elevation, but solid in their structure, were thickly covered with it.

of this huge and lumpish mountain, other more lofty ones, and with more rugged summits, but almost of a black color, came in sight. On reaching the bottom of a steep hill, we entered a small and fertile valley, the fertility of which was the more apparent and the more pleasant from its being shut in, almost on every side, by these high black mountains. At one extremity of this valley, upon an eminence of lava, we remarked several conical masses of rock, which appeared to be the apertures of extinguished craters, and exactly of the same nature as the one we had just left. They, however, were too far from us to allow of our examining them, as it would have detained us a day more, before we could arrive at the Geysers. I therefore proposed staying here, if possible, on my return, and contented myself, for the present, with going a little way up a gulley, in one of the mountains, to look at a cave, which an Icelander in our party had assured us was worth seeing, though I must confess I found in it nothing remarkable. It was an opening in the side of the mountain, barely six feet high, by twenty or thirty feet deep, excavated in a black sand stone, which, (at least that part of it that had not been exposed to the air,) was of a very shining quality. Although the whole of this mountain appeared to be composed of sand-stone rock, yet it was not all equally soft: some lay in interrupted, but horizontal, strata of several feet in thickness, and of a very firm and compact nature, not being so easily washed down by the torrents of snow water, as the rest of the mountain, but remaining firm, and projecting from its sides in various places, and of a browner color. Continuing our journey, we crossed a rugged moor of considerable extent, and at length entered upon an immense plain, a great part of which was either a morass, or covered with a Lake, called Apn-Vatn. From the water near the margin we saw at a distance, at a place known by the name of Laugardalr *, a great quantity of steam rising in three or four columns. On approaching, we found it caused by some boiling-springs, one of which was of considerable size, and proceeded from an opening in the rock in a very shallow part of the lake, throwing up a very beautiful jet about four

* *Laugar* is a term applied to the warm baths, in Iceland.

feet in height, and of nearly the same width. At the margin of the water, nearest the hot-spring, was a border of sulphur, which covered the stones with a thin yellow incrustation. Three or four other boiling-springs, also, were close by, some a little way in the lake, and others rising from the dry ground, but all of a small size. The rest of our road to Middalr, where we proposed passing the night, was along the margin of the lake, and we reached the place about eight o'clock in the evening; having travelled the whole day without resting the horses. Our tents were placed near the church and the house of the priest, who soon came down to welcome us, and offer any thing that his parsonage would afford. As the most necessary, I first requested that we might have some fire prepared to cook our victuals by; during which operation I was witness to a scene that afforded me no small degree of amusement. After Jacob had been gone into the house some considerable time with the fish that was to be dressed for our dinners, I began to be rather impatient, and begged to be shewn into the kitchen, that I might see if any thing had happened. I was conducted thither by a female, who took hold of my hand, and led me through a dark passage and a bed-room, where but a small portion of light was admitted from an aperture in the roof, into the cooking-room, whence so much smoke was rushing out through the sleeping-room, as the only vent, that I hesitated about proceeding, till I found myself dragged in. I with difficulty discovered two or three filthy females sitting on the ground, or on some broken chests, and in the middle of them Jacob on the bare earth. A fire was also on the ground between his legs, over which he held some fish cut in slices, in the fryingpan, an article which caused considerable astonishment among the women. Close by him sat a pretty Icelandic girl, who had won Jacob's regards so much that he every now and then, with his knife, turned out a slice of the fish for her; while she, in return for every piece thus offered, rose from the ground, hugged him about the neck and kissed him. This innocent custom, in use both among the male and female Icelanders, upon the most trivial occasions, was here exemplified in a very strong and ludicrous manner, and so occupied the attention of

Jacob, (who, probably, mistook for a mark of affection, what was in reality nothing more than an expression of gratitude,) that I was obliged to tap the honest fellow on the shoulder, and remind him that I had not yet had my dinner, and that I wished to have some of the fish saved for me. Before going out of the house I was anxious to make some trifling present to the mistress of it, a little, dirty, ugly, old woman, by no means free from cutaneous diseases. I presented to her a snuff-box; but her modesty would at first only allow her to suppose that I meant the contents of it for her. As soon, however, as she was made to understand that the box, also, was to be included in the gift, I had the mortification to find myself, before I was aware of it, in the embraces of this grateful old lady, from which I extricated myself with all possible haste, and performed a most copious ablution at the nearest stream. Of the poverty of the clergy, as well as of the common people in Iceland, I had heard much previously to my coming to Middalr, yet was scarcely prepared for what I here met with, though I had been assured by the priest Egclosen that instances were not wanting of gentlemen of his profession having been reduced in bad winters to such a state, for want of the necessaries of life, that they have been obliged to beg a scanty subsistence from house to house; till, through cold and weakness and hunger, they have perished miserably among the mountains. Their salaries are, usually, exceedingly small: that of the priest of Middalr was only twenty rix-dollars a year, four of which he received from the king. It is true, he added some little to his income by exercising the trade of a blacksmith, but the wretched maintenance which these two professions, so incompatible, happily, in the ideas of an Englishman, conjointly afforded, may be easily conceived, when I mention, that I observed both him and two or three persons of his family eagerly picking up from the ground the heads and entrails of the fish, which Jacob, in preparing for cooking, had thrown away. After dinner, the priest brought down to my tent a present of a large quantity of the *Lichen islandicus (Fiallagros Isl.)*. It is, perhaps, in no country found in such plenty, as in this from which it takes its name. The extensive desert tracts of Skaptar-fel Syssel produce this plant in extreme abundance, and numerous parties from great distances migrate thither, with their horses, tents, and provisions in the summer months, and remain some time, for the sole purpose of gathering it. They then convey it on their horses to Reikevig, or any other factory, and dispose of it to the Danish merchants. Povelsen and Olafsen observe, that a person can collect four tons or a horse-load in a week, and that a peasant is better off with this quantity of the lichen, than with one ton of meal *. It is said to require three years before it has

* Kerguelen, in the *Account of his Voyage to the North*, gives us an extract from a letter of an Icelander, (Mr. Olave) whom he met with in Patrixfiord, where the qualities of this lichen are very highly extolled; perhaps more so than they deserve. " I send to you, Sir, (says Mr. Olave) a herb, which, resembling lung's-wort, serves among the Icelanders as a succedaneum for bread; it is called Iceland-moss, and grows on the rocks of the loftiest mountains; so that with truth we may say, God gives us bread from stones. It never grows in earth or soil of any description, nor casts forth roots. It affords a noble feast; the powder of it is taken in milk, and is so pleasant and salubrious, that I prefer it to every kind of flour; it is, besides, an excellent stomachic, and a most safe medicine in a dysentery."—Amongst many

arrived at its full growth; for, having once cleared a spot of ground by gathering the lichen, the natives wait always that length of time before they visit the same place

other good qualities of the mind which Icelanders in general possess, contentment with the station in which Providence has placed them, and a strong sense of gratitude for the supplies which the Deity is pleased to grant to them, are, certainly, the most predominant. Cut off by the situation and poverty of his native land from almost all communication with happier climates, where plenty and luxury abound, an Icelander is ignorant even of their existence, and eats his dried uncooked fish, and rancid butter with a grateful heart. He possesses the *amor patriae* in as strong a degree as the inhabitant of any country. Volcanoes, which have laid waste his whole island, earthquakes, disease, and famine, cannot drive him from his native shores. The few who have gone over to Denmark have expressed the greatest desire to return home, although the kindest treatment, and every thing that was likely to make them comfortable, had been employed to induce them to remain. The man, who was my guide during most of my excursions in Iceland, had himself passed two years in Copenhagen, and, although, as he confessed to me, it was a milder climate and he had better living in Copenhagen, yet he had much rather spend his days where he then was. Besides the *Lichen islandicus,* Povelsen and Olafsen notice three other species of Lichen which are occasionally caten. *Lichen proboscideus,* the *Coralloides tenuissimum nigricans* of Dill., and *L. nivalis.* This

again, when they find another harvest. The only necessary preparation previous to cooking is to steep the lichen in clean cold water for some time, for the purpose of extracting the strong bitter taste which is peculiar to it: it is then dried in the sun, reduced to powder, and boiled up with milk, till it has become of such a consistency as to be quite a jelly when cold. As an article of food it is commonly eaten, dressed in the above-mentioned way, and is considered both very wholesome and nourishing; nor does it by any means possess that purgative quality, which Linnæus and others have attributed to it; but which may exist in the bitter that has been previously extracted by the steeping in water. I do not think its medicinal virtues, in pulmonary complaints, for which it is so highly valued in other countries, are at all generally known to the Icelanders, many of whom expressed great surprise when I mentioned to them the circumstance. The good old priest, after having presented his Ice-

latter, which grows in great quantity about Reikevig, is called by the natives, *Maringraus*, or the *Virgin Mary's Grass*, and is said to be extremely agreeable food, and of a sweet taste.

land-moss, requested some medical advice, supposing from my fondness for plants that I must be a physician. In this I was sorry to be obliged to undeceive him, and, indeed, I could only do it with great difficulty. He wished me much to examine his hip, which had been some years ago dislocated, and had healed very awkwardly for want of surgical assistance. A wound, also, which he received at the same period, had ulcerated, and he had been able to procure no application since that time twelvemonth, when, as he said, a gentleman, with a star upon his breast, gave him a plaister. He was travelling to the Geysers, but who he was he could not tell. When I at length assured him that it was not in my power to render him any service, his wife's diseases were enumerated, and I was entreated to examine her sores. On my declining this, he resolved to turn physician himself, and begged me to give him some rum to bathe his wife's breast: to this I consented; but, after having applied a portion of it to that purpose, he drank the rest, without being at all aware of its strength, which, however, had no other effect than the very ludicrous one of causing this clerical blacksmith with his lame hip to dance, in the most ridiculous manner, in the front of the house. The scene afforded a great source of merriment to all his family, except his old wife, who was very desirous of getting him to bed, while he was no less anxious that she should join him in the dance. The wife, however, at length gained the victory, and he retired in great good humour *.

* I should be extremely sorry, if, by this little anecdote, I am supposed to intimate that drinking is a common vice among the Icelanders. I have every reason to think very much the contrary. Indeed, this very circumstance is a convincing proof how unaccustomed the priest of Middalr was to spirituous liquors: otherwise, the small quantity he drank, which could not at any rate have exceeded a wine-glass full, would not have elated his spirits so much. At Reikevig, it is true, drunkenness, and almost every other vice, have been introduced by the Danes, but they are confined solely to the town, and principally to the Danes themselves. I do not recollect, during the whole of my stay in the island, that I saw half a dozen natives much in liquor, and those were all in Reikevig. Their morals are extremely correct. It is not without the most thorough contempt for the author of such a falsehood, that I read the following passage, extracted from *Anderson's History of Iceland:* "These people know very little of God, or his will; for the value of two marks, or sixteen-pence, they will perjure themselves even to the prejudice of their nearest

Thursday, July 13. This morning we had rain and squalls. After breakfast the priest came down, and begged that he might be allowed to accompany me to the Geysers; but this I could by no means consent to, as it was my full intention to proceed to Hecla, and to return by another route. He insisted, however, upon conducting me some way on my road, and especially across a river, which

relations; full of wrath and revenge, extremely lascivious and vicious, and errant thieves and cheats. What, then, can be expected from a people that have no awe or check, and live in an unbridled licentiousness, without any restraint ashore and at sea, frequent opportunities unobserved, and consequently unpunishable, and continually indulging themselves in the filthy sin of drunkenness?"—These absurd falsities are scarcely deserving of refutation. Were such conduct, as is here mentioned, really to exist, it could not but be productive of the most serious consequences to the nation: the prison-houses would be filled with persons, who would have been gradually led on to commit the worst of crimes. Yet, that such is not the case in Iceland, may be believed, when it is known that there is only one prison for 48,000 inhabitants, and that, on our arrival, which was a little previous to a sitting of the court of justice, there was only one criminal in it (and even this was more than had been the case for a long time), and five or six persons confined for small offences.

he called Brueraa, and which, owing to the late wet weather, he thought might probably be too deep to cross to-day. He accordingly went to his wardrobe in the church, dressed himself in his best clothes, and was ready to start with us. We continued our journey along the foot of a barren mountain, at no great distance from the marshes. Here and there, indeed, we met with a few stunted birch-trees, but no plants that I had not seen elsewhere. Leaving the mountain, and crossing a disagreeable swamp, we, in about two or three hours, arrived at the most fordable part of the Brueraa. There was already a party of horsemen there, resting their horses a little, to prepare them for the fatigue of passing through this stream, the bottom of which is exceedingly rocky, and the river itself both wide and deep, but at this time considered fordable. The packages of fish, wool, &c., were carefully fixed by ropes to the top of the horses' backs, so that they might be as little exposed to the water as possible; and the horses, being then tied in a line one behind the other, all reached the opposite shore in safety, though the smaller ones were compelled to swim. A foal, which was fastened by the neck to the tail of its mother, was dragged through, and landed on the other side of the river, more dead than alive, through fear and cold. Our party followed, and was equally fortunate in getting over without any accident (except the wetting of the luggage and ourselves), though the water reached to the middle of the body of our tallest horses. Here, after procuring us some milk from a cottage close by, the priest took his leave of us. In the vicinity of the house were two or three boiling-springs, which were used by the inhabitants for the purpose of cooking, as well as for that of washing their clothes. At a few miles distance, on our right, we saw a very considerable column of steam, rising from the marshes, at a place which the guides called Reykum *, and which they said I might visit on my way to Skalholt. Our journey now lay either entirely over a morass, which proved extremely

* This is not the *Reykum*, or *Rykum*, which Sir John Stanley has given so full and so admirable an account of: many places are called by this and similar names, derived from the Icelandic word *Reik*, or *Reyk*, which signifies *smoke*; such are *Reykholt*, *Reikevig*, *Reikholtsdal*, *Reikanaes*, &c.

fatiguing to our horses, or upon the edge of it, where a quantity of loose soil had been washed down from the mountains by the torrents, and was scarcely more firm than the bog itself. At about five o'clock in the afternoon we obtained the first view of the mountain, called Laugerfell, from which the Geysers spring. It is of no great elevation, rising, according to Sir John Stanley, who had an opportunity of ascertaining by admeasurement, only three hundred and ten feet above the course of a river which runs at its foot. It is, however, remarkable for its insulated situation; being entirely surrounded by a morass, which extends for a very considerable way in every direction, except towards the north, where this hill is not separated by an interval of more than half a mile from higher mountains. The north side is perpendicular, barren, and craggy; the opposite one rises with a tolerably gradual ascent, and from this, near its base, we saw a number of columns of steam mounting to various heights. Enlivened by the prospect, we quickened our pace, and at eight o'clock arrived at the foot of the hill. Here I left my horses, &c., to the care of the guides, and hastened among the boiling-springs, happy in the prospect of soon beholding what may justly be considered as one of the most extraordinary operations of nature, and thus accomplishing one of my principal objects in undertaking a voyage to Iceland. The lower part of the hill was formed into a number of mounds, composed of what appeared to be clay or coarse bolus, of various sizes: some of them were yellowish white, but the greater number of the color of dull red brick. Interspersed with them, here and there, lay pieces of rock, which had rolled, or been washed down by the rains, from the higher parts of the mountain. On these mounds, at irregular distances, and on all sides of me, were the apertures of boiling-springs, from some of which were issuing spouts of water, from one to four feet in height; while in others, the water rose no higher than the top of the basin, or was gently flowing over its margin. The orifices were of various dimensions, some of considerable size and regular formation, covered on their sides and edges with a brownish siliceous crust; others so small and irregular that the water seemed only to be boiling through an accidental hole in the mound, and became turbid by admixture with the soil, which colored it either with red, dirty yellow, or grey. Upon the heated ground, in many places, were some extremely beautiful, though small, specimens of sulphuric efflorescence, the friability of which was such, that, in spite of the utmost care, I was not capable of preserving any in a good state. I did not remain long in this spot, but directed my steps to the loftiest column of steam, which I naturally concluded arose from the fountain that is alone, by way of distinction, called *the Geyser*. It lies at the opposite extremity of this collection of springs, and, I should think, full half a quarter of a mile distant from the outermost ones which I first arrived at. Among numerous smaller ones, I passed three or four apertures of rather a large size, but all so much inferior to the one I was now approaching, that they scarcely need any farther notice. It was impossible, after having read the admirable descriptions of the Geyser, given by the Archbishop Von Troil and Sir John Stanley*, and, especially,

* I need scarcely refer my readers for a more full account of the Geyser than it is in my power to give, to the letters of Von Troil, who accompanied Sir Joseph

after having seen the engravings made from drawings taken by the last-mentioned gentleman, to mistake it. A vast circular mound, (of a substance which, I believe, was first ascertained to be siliceous by Professor Bergman,) was elevated a considerable height above those that surrounded most of the other springs. It was of a brownish grey color, made rugged on its exterior, but more especially near the margin of the basin, by numerous hillocks of the same siliceous substance, varying in size, but generally about as large as a mole-hill, their surface rough with minute tubercles, and covered all over with a most beautiful kind of efflorescence; so that the appearance of these hillocks has been aptly compared to that of the head of a cauliflower. On reaching the top of this siliceous mound, I looked into

Banks in his voyage to Staffa and Iceland: the work is too well known to every one. The two excellent letters of Sir John Stanley on the hot-springs near Rykum, and on those near Haukardal, are to be found in the third volume of the *Transactions of the Society of Edinburgh.* In the same volume, also, is to be met with a full account of the analysis of the water of the hot-springs, by the late Dr. Black, of Edinburgh.

the perfectly circular basin*, which gradually shelved down to the mouth of the pipe or crater in the centre, whence the water issued. This mouth lay about four or five feet below the edge of the basin, and proved, on my afterwards measuring it, to be as nearly as possible seventeen feet distant from it on every side; the greatest difference in the distance not being more than a foot. The inside was not rugged, like the outside; but apparently even, although rough to the touch, like a coarse file: it wholly wanted the little hillocks and the efflorescence of the exterior, and was merely covered with innumerable small tubercles, which, of themselves, were in many places rendered quite smooth and polished by the falling of the water upon them. It was not possible now to enter the basin, for it was filled nearly to the edge with water the most pellucid I ever beheld, in the centre of which was observable a slight ebullition, and a large, but not dense, body of steam, which, however, increased both in quantity

* To compare great things with small, the shape of this basin resembles that of a saucer with a round hole in its middle.

and density from time to time, as often as the ebullition was more violent. At nine o'clock I heard a hollow subterraneous noise, which was thrice repeated in the course of a few moments; the two last reports following each other more quickly than the first and second had done. It exactly resembled the distant firing of cannon, and was accompanied each time with a perceptible, though very slight, shaking of the earth; immediately after which, the boiling of the water increased together with the steam, and the whole was violently agitated. At first, the water only rolled without much noise over the edge of the basin, but this was almost instantly followed by a jet*, which did not rise above ten or twelve feet, and merely forced up the water in the centre of the basin, but was attended with a loud roaring explosion: this jet fell as soon as it had reached its greatest

* I have followed Sir John Stanley in using the word *jet* for this sudden shooting of the water into the air, which continues but a few seconds, because I do not know that we have any term more applicable in our language. The French employ the word *élancement* in the same sense, which seems to convey a better idea of the thing, but cannot well be rendered in English.

height, and then the water flowed over the margin still more than before, and in less than half a minute a second jet was thrown up in a similar manner to the former. Another overflowing of the water succeeded, after which it immediately rushed down about three-fourths of the way into the basin. This was the only discharge of the Geyser that happened this evening. Some one or other of the springs near us was continually boiling; but none was sufficiently remarkable to take off my attention from the Geyser, by the side of which I remained nearly the whole night, in anxious but vain expectation of witnessing more eruptions. It was observed to us by an old woman, who lives in a cottage at a short distance from the hot-springs, that the eruptions of the Geyser are much most frequent, when there is a clear and dry atmosphere, which generally attends a northerly wind; and we now congratulated ourselves upon the prospect of being enabled to ascertain the accuracy of her observation, the wind, which had hitherto continued to the south-west, having this evening veered about to the north. At twenty minutes past eleven

Friday, July 14. on the following morning, I was apprised of an approaching eruption by subterraneous noises and shocks of the ground, similar to those which I had heard and felt the preceding day; but the noises were repeated several times, and at uncertain, though quickly recurring, intervals. I could only compare them to the distant firing from a fleet of ships on a rejoicing day, when the cannon are discharged without regularity, now singly, and now two or three almost at the same moment. I was standing at the time on the brink of the basin, but was soon obliged to retire a few steps by the heaving of the water in the middle, and the consequent flowing of its agitated surface over the margin, which happened three separate times in about as many minutes. A few seconds only had elapsed, when the first jet took place, and this had scarcely subsided before it was succeeded by a second, and then by a third, which last was by far the most magnificent, rising in a body that appeared to us to reach not less than ninety feet in height, and to be in its lower part nearly as wide as the basin itself,

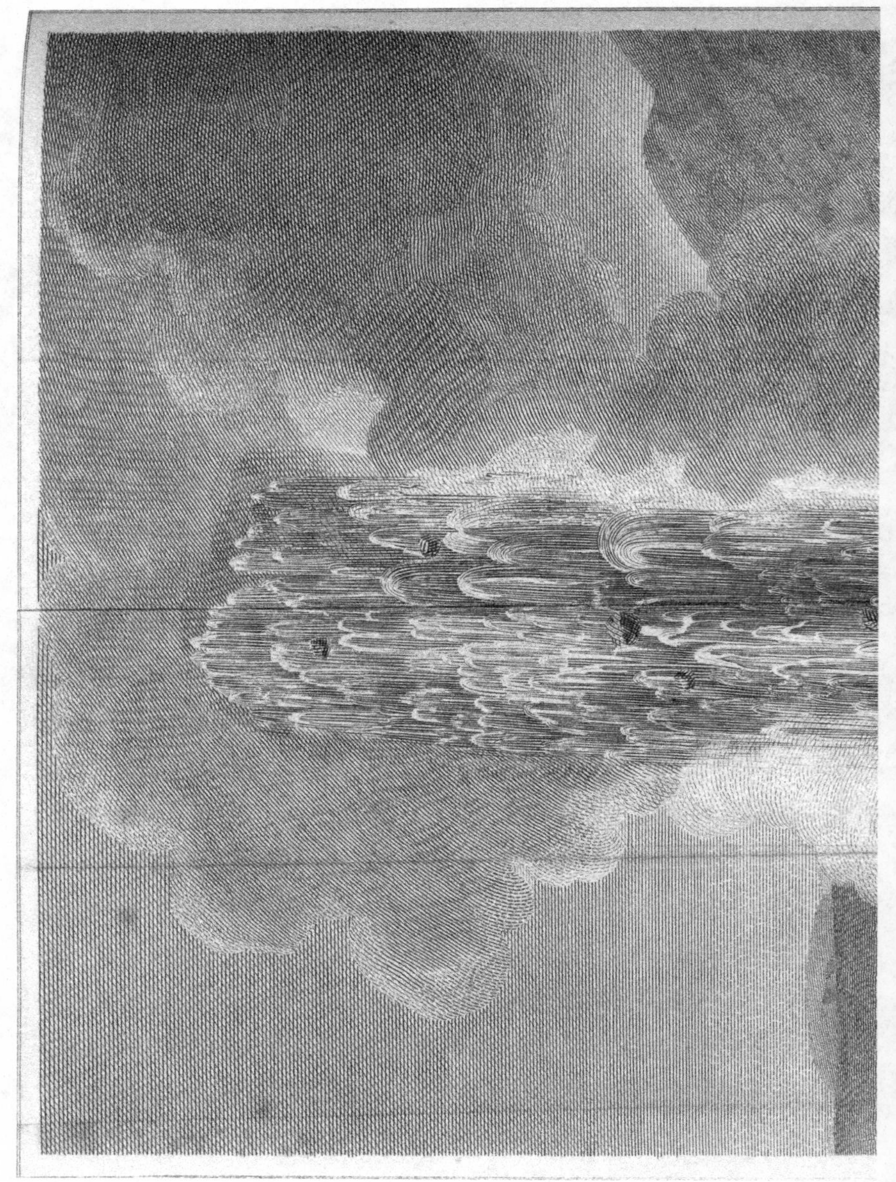

Engraved by W. H. C. Edwards, from Sir Joseph Banks' Collection of Drawings.

Eruption of the Geyser.

which is fifty-one feet in diameter. The bottom of it was a prodigious body of white foam, magnificent beyond what the warmest imagination could picture, and by concealment rendering more impressive the wonders it envelopped; but, higher up, amidst the vast clouds of steam that had burst from the pipe, the water was at intervals discoverable, mounting in a compact column, which at a still greater elevation, where it was full in view, burst into innumerable long and narrow streamlets of spray, some of which were shot to a vast height in the air in a perpendicular direction, while others were thrown out from the side, diagonally, to a prodigious distance *. The excessive transparency of the body of water, and the brilliancy of the

* Darwin, in his *Botanic Garden*, vol. i. page 128, has a few lines upon the Geyser, which are rather more poetical than correct:

High in the frozen north where Hecla glows,
And melts in torrents his coeval snows;
O'er isles and oceans sheds a sanguine light,
And shoots red stars amid the ebon night;
When, at his base entombed, with bellowing sound
Fell Geyser roar'd, and, struggling, shook the ground;

drops, as the sun shone through them, considerably added to the beauty of the spectacle. As soon as the fourth jet was thrown out, which was much less than the former,

Pour'd from red nostrils, with her scalding breath,
A boiling deluge o'er the the blasted heath;
And wide in air its misty volumes hurl'd
Contagious atoms o'er the alarmed world:
Nymphs, your bold myriads broke the infernal spell,
And crush'd the sorceress in her flinty cell."

In these two last lines the Doctor alludes, as he tells us in a note, to the eruption of a volcano, which happened subsequently to the time of Sir Joseph Banks' being there, and which extended as far as the Geysers, and overflowed them with its lava. Whence he could have obtained this piece of information, I am at a loss to guess: certainly it was not from any book of good authority, for no such circumstance has happened.—This reminds me of a similar error in *Dr. Adam's Geography*, where it is said that Hecla is constantly spouting out fire and hot water; and, with regard to the religion of the Icelanders, that most of them are Lutherans, but that there are some Pagans. The Etatsroed, who possesses a very mild temper, which I never saw ruffled, even in trying circumstances, was still unable to restrain himself when he pointed out these inaccuracies to me, and denied the veracity of them with considerable warmth, quoting passages from English authors who had written previously to the time of Doctor Adam,

View of the Crater of Geyser when empty immediately after an Eruption.

and scarcely at the interval of two minutes from the first, the water sunk rapidly in the basin, with a rushing noise, and nothing was to be seen but the column of steam, which had been continually increasing from the commencement of the eruption, and was now ascending perpendicularly to an amazing height, as there was scarcely any wind, expanding in bulk as it rose, but proportionably decreasing in density, till its upper part gradually lost itself in the surrounding atmosphere. I could now walk in the basin to the margin of the pipe, down which the water had sunk about ten feet, but it still boiled, and every now and then, furiously and with a great noise, rose a few feet higher in the pipe, then again subsided, and remained for a short time quiet. This continued to be the case for some hours. I measured the pipe, and found it to be exactly

and who had stated the facts as they really were. He begged me, on my return, to make Doctor Adam acquainted with the incorrectness of his remarks upon Iceland, that they might be altered in a future edition of his work.—But the time is past, for the worthy Doctor is dead!

seventeen feet over, and, as I have before mentioned, situated in the very centre of the basin, which was fifty-one feet in diameter. The pipe opens into the basin with a widened mouth, and then gradually contracts for about two or three feet, where it becomes quite cylindrical, and descends vertically to the depth, according to Povelsen and Olafsen, of between fifty and sixty feet. Its sides are smooth, and covered with the same siliceous incrustation as the basin. It was full twenty minutes after the sinking of the water from the basin, before I was able to sit down in it, or to bear my hands upon it without burning myself. At half past two o'clock it was again nearly filled, the water having risen gradually, but at intervals, attended every now and then with a sudden jet, which, however, did not throw it more than two or three feet higher than the rim of the basin. A few minutes after, there was a slight eruption, but the greatest elevation to which the water was ejected, was not above twelve feet. At four o'clock in the afternoon my guide was witness to another, while I was away. I had been visiting the other hot-springs, and, amongst them, that which Sir John Stanley calls *the Roaring Geyser*, in which, though the water rose and fell several feet at uncertain intervals, and was frequently boiling with a loud and roaring noise, I still did not perceive that it ever flowed over the margin of the aperture. Its pipe or well does not descend perpendicularly, but, after going down some way in a sloping direction, seems to continue in a nearly horizontal course. Around its mouth lies a considerable quantity of red earth, or bolus, and on one side of it I observed, what appeared to me, a curious mineralogical production: it was imbedded in a hard kind of rock, but was of itself exceedingly brittle, and apparently fibrous; looking much like asbestos, but materially differing from that mineral in its extremely fragile nature. On going to the foot of the hill, near the spot where the waters of the Geyser join a cold stream, among the numerous rills which the heated water had formed, I met with some uncommonly beautiful specimens of incrustations. Every blade of grass, and every leaf or moss that was washed by these waters, was clothed with a thin covering of the same siliceous substance as the great basin was composed of, but of so delicate a nature that it was scarcely possible, even with the utmost care, to bring any of them away perfect. I remarked, in particular, a *Jungermannia (asplenioides)* so beautifully coated with this incrustation, that it looked as if it were a model of the plant in plaster of Paris. One specimen was so protected under the shelter of larger plants incrusted together, that I was able to convey it in safety to Reikevig. The plants I met with by the side of the river, which I had not remarked before, were *Carex Bellardi* and a new species of the same genus, with *Koenigia islandica* in great profusion, and *Funaria hygrometrica*. Leaving the river, I walked over several vast mounds of red earth, at the north end of the Geyser, in my way to the top of the mountain. Here and there a boiling-spring was forcing its turbid and discolored waters through holes in the surface. Some were completely in the thick muddy state of a puddle, and were bubbling, as any glutinous substance would do over a fire. In many places was heard a rumbling noise like the subterraneous boiling of water, although there was no orifice near, by which the fluid could make its escape. On these spots, which were so much heated by the boiling streams beneath that I could scarcely bear my hands upon the ground, I found a great profusion of *Riccia glauca* *, growing in patches, and extending almost uninterruptedly over a space of ten or twelve feet in diameter. The soil for more than half way up the mountain was composed of a coarse reddish kind of earth, intermixed with some other of a dirty yellow color, with small intervals of hard rock, and with this terminated the highest of the hot-springs, which, however, was but a feeble one. Thence to the summit the mountain was entirely formed of a loosely-laminated rock, whose strata seemed to lie in almost every direction, but chiefly vertically. There was no appearance whatever of any part of the hill having been in a state of fusion. Many of the strata were still in their ori-

* I think, but dare not trust too implicitly to my memory, that I saw abundance of it in fructification. I made no memorandum on this subject, and the specimens which were intended to enable me to answer this, as well as other questions relative to natural history, were all, unhappily, lost.

ginal bed, and the pieces which had fallen from them had their edges very sharply defined, and had broken off in laminæ, of about an inch in thickness. The stone is extremely hard and compact, of a rusty brown color, in some specimens more inclining to grey, and with a perfectly smooth and flat surface. Sir John Stanley supposes that its substance is chiefly argillaceous, and that, like every other stone in the island, it has undergone some change by fire. I met with nothing remarkable on the summit, where there is a considerable extent of flat surface, almost covered with *Trichostomum canescens,* intermixed with the *Lichen islandicus;* and from each extremity of this plain arises a conical eminence, unequal in height, of the same nature as the rock it springs from, and producing no plants that are not to be seen equally abundant in various other parts of the country. The most scarce were *Trichostomum ellipticum,* which grows in tolerable plenty upon the dry rocks, and *Andraea Rothii,* which, though it has been found in but few countries, is very abundant in Iceland. The top of Laugerfell afforded me a very commanding prospect.

Just beneath me, facing the south-east, was to be seen, at one view, the steam rising from upwards of a hundred boiling-springs, among which the great Geyser, from its regularly circular figure, looked like an artificial reservoir of water. A little stream at the bottom of the hill formed the boundary to these, and beyond this was an extensive morass, whose sameness was only interrupted by the rather wide course of the river Hvitaa, winding through it. The view was terminated, in that quarter of the compass, by a long range of flat and tame mountains, over which towered the three-pointed and snow-capped summit of Hecla, which rises far above the neighboring hills, and is, in clear weather, plainly visible when standing by the Geyser. In the north-east was situated the church and farm of Haukardal, and a continuation of the morass, bounded by some lofty jökuls of fantastic shapes. In the north-west, at a small distance from the place where I stood, and, indeed, only separated from it by a narrow portion of the morass, with a small river winding through it, rose another chain of mountains, thinly covered with vegetation, beyond which some jökuls shewed their white summits. In the south the morass was extended almost to the coast, and looked like a great sea, having three or four rather lofty, but completely insulated, mountains, with flat summits, rising from its bosom. It was my custom, during my stay in this place, to cook my provisions in one or other of the boiling-springs; and, accordingly, a quarter of a sheep was this day put into the Geyser, and Jacob left to watch it, holding it fastened to a piece of cord, so that, as often as it was thrown out by the force of the water (which very frequently happened), he might readily and without difficulty drag it in again. The poor fellow, who was unacquainted with the nature of these springs, was a good deal surprised, when, at the time he thought the meat nearly cooked sufficiently, he observed the water in an instant sink down, and entirely disappear; not rising again till towards evening. This disappointment therefore obliged us to have recourse to another spring, and we found, that, in all, it required twenty minutes to perform the operation properly. It must be remembered, however, that the quarter of an Icelandic sheep is very small, perhaps not weighing more than six pounds, and is, moreover, extremely lean. I do not apprehend that longer time would have been necessary to have cooked it in an English kitchen; for the hot-springs in Iceland, at least such of their waters as are exposed to the air, are never of a greater heat than 212° of Fahrenheit: so that, when I hear travellers speaking of having boiled their eggs in two minutes in such springs, or of having cooked their meat in a proportionably short space of time, without presuming to doubt the fact, I must be allowed to suspect that their victuals would not have been dressed to my taste. The next eruption of the Geyser, which took place at half past nine, was a very magnificent one, and was preceded by more numerous shocks of the ground and louder subterraneous noises, than I had yet witnessed. The whole height to which the greatest jet reached could not be so little as a hundred feet. It must be observed, however, that I had no instruments with me for measuring elevations, and therefore, could only judge by my eye; Jacob and myself watching at the same time, and each giving his estimate. The difference between us was but trifling, and I always took the lowest calculation. My method was, to compare the height of the water with the diameter of the basin, which I knew to be fifty-one feet, and this jet was full twice that length. The width of the stream is not equally easily determined by the eye, on account of the steam and spray that envelops it: in most instances, not more, probably, than eighteen or twenty feet of the surface of the water is cast into the air; but it occasionally happens, as was the case now, that the whole mass, nearly to the edge of the basin, is at once heaved up: all, however, is not spouted to an equal height; for the central part rises the highest, but, having gained some elevation, the spray divides, and darts out little jets on every side, that fall some way over the margin of the basin. After this last discharge, the water subsided about fifteen feet in the pipe, and so remained some time; but in two hours the funnel was again filled to within two feet of the edge. As often as I tried the heat of the water in the pipe, I always found it to be 212°; but, when the basin also was full, on immersing the thermometer as far from the margin as I could reach with my arm, I never saw the quicksilver rise above 180°; although the water in the centre was boiling at the same time. It seems probable that the height to which the Geyser throws its waters may have increased somewhat in the course of a few years; as, when Sir Joseph Banks visited Iceland in 1772, the greatest elevation to which the column ascended, was ascertained to be ninety two feet; while in the year 1789, its height was taken by a quadrant, by Sir John Stanley, and found to be near one hundred feet, and this day, if I am not mistaken, it was still greater. Povelsen and Olafsen were in all probability deceived, when they imagined they saw the loftiest jets reach to the elevation of sixty toises, or three hundred and sixty feet. Previously to the last eruption, Jacob and myself amused ourselves with throwing into the pipe a number of large pieces of rock and tufts of grass, with masses of earth about the roots, and we had the satisfaction to find them all cast out at the discharge, when many of them fell ten or fifteen feet beyond the margin. Some rose considerably higher than the jets which forced them up: others fell down into the basin, and were with the following eruption a second time flung out. The stones were mostly as entire as when they were put in, but the tufts of grass and earth were shivered into numerous small black particles, and were thrown up by the first jet in quick succession, producing a very pleasing effect among the white spray. This whole day had been fine with but little rain.

Saturday, July 15. At one o'clock this morning there was an eruption of the Geyser, which was repeated at half past three, and again at a quarter before eight, and at half past nine; after which, the fountain continued to spout water about every two hours. All these eruptions were attended by the same circumstances as those of yesterday, and were preceded by similar tremblings of the ground and subterraneous noises; but none of them threw the water to any great elevation; the highest not appearing to exceed fifty feet. Close to the edge of many of the hot-springs, and within a few inches of the boiling water, in places that are, consequently, always exposed to a considerable degree of heat, arising both from the water itself and the steam, I found *Conferva limosa Dillw.* in abundance, forming large dark-green patches, which easily separated and peeled off from the coarse white kind of bolus that they were attached to. In a similar situation, also, I met with a new species of *Conferva* (belonging to the family called by Vaucher *Oscillatoria*), of a brick-red color, covering several inches of ground together, and composed of extremely minute unbranched filaments, in which, with the highest powers of my microscope, I was not able to discover any dissepiments. The margin of one of the hot-springs, upon a white bolus, which was in a state of puddle from its mixture with the heated water, afforded me the finest specimens of *Jungermannia angulosa* * I ever saw, growing thickly matted in such great

* Mr. Barrow, in his *Voyage to Cochin-China*, gives us a very interesting account of the hot-springs in the island of Amsterdam, which lies in latitude 38° 42′ south, and longitude 76° 51′ east. "Some of them," he says, "are running freely, others ooze out in a paste or mud. In some of the springs Fahrenheit's thermometer ascended from 62° in the open air to 196°; in some to 204°; and in others to 212°, or the boiling point. In several places we observed patches of soft verdure, composed of a fine delicate moss, blended with a species of *Lycopodium* and another of *Marchantia*. These green

tufts, that I could with ease take off pieces of five or six inches in diameter. The under side of these patches had very much the appearance of purple velvet, owing to the numerous fibrous radicles of that color which proceeded from the base of the stems, and suffered themselves to be detached, without difficulty, from the soil they had grown upon. In water, also, of a very great degree of heat, were, both abundant and luxurious, *Conferva flavescens* of Roth, and a new species allied to *C. rivularis.* After a day, almost the whole of which had been showery, with the wind in the south-west, a fine,

Sunday, July 16.

but cold, morning, attended with a northerly wind, afforded me a most interesting spectacle, the idea of which is too strongly impressed on my mind, ever to be obliterated but with memory itself. My tent had been pitched at the distance of

patches were found to be floating on a hot paste, whose temperature, at eight or ten inches below the surface, upon which the roots of the plants spread, was 186°. This was the more remarkable, as the same species of *Lycopodium*, or club-moss, grows with great luxuriance, even in the winter season, on the black heaths of North Britain."

three or four hundred yards from the Geyser, near a pipe or crater of considerable dimensions, in which I had hitherto observed nothing extraordinary. The water had been almost constantly boiling in it, and flowing gently over the mouth, thus forming a regular channel, which, I believe, had never ceased running during the whole time of my stay. My guide, however, had informed me that sometimes the eruptions of this spring were very violent, and even more remarkable than those of the Geyser, and it was on this account that he had placed the tents so close to it. At half past nine, whilst I was employed in examining some plants gathered the day before, I was surprised by a tremendously loud and rushing noise, like that arising from the fall of a great cascade immediately at my feet. On putting aside the canvass of my tent, to observe what could have occasioned it, I saw within a hundred yards of me a column of water rising perpendicularly into the air, from the place just mentioned, to a vast height; but what this height actually was I could form no idea; and so overpowered was I by my feelings, that I did not, for some time, think of endeavoring to ascertain it. In my first impulse I hastened only to look for my portfolio, that I might attempt, at least, to represent upon paper what no words could possibly give an adequate idea of; but in this I found myself nearly as much at a loss as if I had taken my pen for the purpose of describing it, and I was obliged to satisfy myself with very little more than the outline and proportional dimensions of this most magnificent fountain. There was, however, sufficient time allowed me to make observations; for, during the space of an hour and half, an uninterrupted column of water was continually spouted out to the elevation of one hundred and fifty feet, with but little variation, and in a body of seventeen feet in its widest diameter; and this was thrown up with such force and rapidity, that the column continued to nearly the very summit as compact in body, and as regular in width and shape, as when it first issued from the pipe; a few feet only of the upper part breaking into spray, which was forced by a light wind on one side, so as to fall upon the ground at the distance of some paces from the aperture. The breeze, also, at times, carried the immense volumes of steam that accompanied the eruption to one side of the pillar of water, which was thus left open to full view, and we could clearly see its base partly surrounded by foam, caused by the waters striking against a projecting piece of rock, near the mouth of the crater; but thence to the upper part, nothing broke the regularly perpendicular line of the sides of the water-spout, and the sun shining upon it rendered it in some points of view of a dazzling brightness. Standing with our backs to the sun, and looking into the mouth of the pipe, we enjoyed the sight of a most brilliant assemblage of all the colors of the rainbow, caused by the decomposition of the solar rays passing through the shower of drops that was falling between us and the crater. After the water had risen to the vast height above described, I ventured to step in the midst of the thickest of the shower of spray; where I remained till my clothes were all wetted through, but still scarcely felt that the water was warmer than my own temperature. On the other side of the spout, so undivided was the column, that, though upon the very brink of the crater, within a few inches of the stream, I was neither wetted nor had I a fear of being scalded by any scattered or falling drops. Stones of the largest size that I could find, and great masses of the siliceous rock, which we threw into the crater, were instantly ejected by the force of the water, and, though the rock was of so solid a nature as to require very hard blows from a large hammer, when I wanted to procure specimens, they were, nevertheless, by the violence of the explosion, shivered into small pieces, and carried up with amazing rapidity to the full height of, and frequently higher than, the summit of the spout. One piece of a light porous stone was flung at least twice as high as the water, and falling in the direction of the column, was met by it, and a second time forced up to a great height in the air. The spring, after having continued for an hour and half spouting its waters in so lofty a column, and with such amazing force, experienced an evident diminution in its strength; and, during the space of the succeeding half hour, the elevation of the spout varied, as we supposed, from twenty to fifty feet; the fountain gradually becoming more and more exhausted, and sometimes remaining completely still for a few minutes, after which it again feebly raised its waters to the height of not more than from two to ten feet, till, at the expiration of two hours and a half from the commencement of the eruption, it ceased to play, and the water sunk into the pipe to the depth of about twenty feet, and there continued to boil for some time. I had no hesitation in pronouncing this to be, what is called by Sir John Stanley, *the New Geyser* *; although the shape and dimensions of the crater differ somewhat from the descriptions given by that gentleman. But, after a lapse of twenty years, it is not to be expected that, with two such powerful agents as fire and water, constantly operating, a spot like this should be suffered to remain without any alteration. The outline of the aperture is an irregular oval, seventeen feet long and nine feet in width; on only one side of which there is a rim or elevated margin, about five

* The term *Geyser*, it may be here remarked, is derived from an Icelandic word which implies a vomiting forth, or boiling out, in a furious manner, and at intervals. "Nomen habet (the learned rector of Skalholt writes to Sir Joseph Banks) a verbo islandico *ad giosa* evomere, ebullire; aquas enim per intervalla in altum evomit."

or six feet in length and one foot high; but the ends of this are ragged, as if it had formerly been continued the whole way round the crater, and it is therefore probably a portion of the same wall, which Sir John Stanley describes as nearly surrounding the basin at the time he was there, and as being two feet high. There is at present no basin whatever round the edge of the pipe, as in the Geyser, nor is the well formed by any means with the same almost mathematical accuracy as in that spring, but on the contrary it is extremely irregular in its figure, and descends in rather a sloping direction; its surface being composed of a siliceous crust, of a deep greyish brown color, worn smooth by the continued friction of the water. These two fountains likewise differ materially in another circumstance, that no subterraneous noises announce the coming eruptions of the New Geyser, or accompany it while it is playing. For several yards, in one direction, in the neighborhood, where the water flows off in a shallow stream, the bed of this is composed of a thin white covering, of a siliceous deposit. During the eruption of the New Geyser, I could not perceive that it in any way affected the neighboring springs. I remarked no particular sinking of the water in any, nor did I observe that any boiled more violently than usual. The Geyser, which was filled almost to the rim of the basin, previously to the eruption of the New Geyser, from which it is distant about four hundred yards or more, remained, as nearly as possible, in the same state of fulness during, and after, the eruption. Sir John Stanley, also, observed the same circumstance, so that in all probability their subterraneous streams are quite independent of each other *. We were informed by the people living in the neighborhood, that in

* Horrebow, indeed, seems to lead to a contrary conclusion, from the following observations: "In the parish of Huusevig, at a farm called Reykum, there are three springs which lie about thirty fathoms from each other. The water boils up in them in the following manner: when the spring or well at one end has thrown up its water, then the middle one begins, which subsiding, that at the other end rises, and after it, the first begins again, and so on in the same order by a continued succession, each boiling up three times in about a quarter of an hour." Page 21.—Povelsen and Olafsen, also, mention a remarkable circumstance, which proves a communication between the two springs, called Akra-

the spring of last year (1808), a violent shock of an earthquake was felt, which made an aperture for another hot-spring, and caused the whole of them to cease flowing for fifteen days. The ground, at that time, appeared to be lifted up some feet; a house was thrown down; and all he cattle, which were at pasture, ran home to the dwellings of their masters, and shewed symptoms of the utmost terror. Earthquakes in this quarter of the country are not unfrequent. One happened but a short time previous to the visit of Sir John Stanley, who conjectures, that this probably enlarged the cavities, communicating with the bottom of the pipe of the New Geyser; for it is to be remarked, that till then (June, 1789), that spring had not played for a considerable length of time with any degree of violence*. A party of horse-

ver, in the canton of Olves, situated at the distance of an hundred toises from each other. On throwing in the lead, for the purpose of sounding the depth of one of these wells, they found the water immediately diminished a foot and a half in depth, whilst at the same time it flowed over from the other well.

* See *Edinburgh Transactions*, v. iii. p. 150.

women †, well dressed, and riding, some astride and some on the saddles of the country, who were passing the Geysers, and directing their course towards Haukardal, reminded me that service was about to be performed at the church of that place this morning, and therefore, as I saw no probability of a second eruption of the New Geyser immediately taking place, I resolved to leave

† These ladies with their long riding coats and their tall head-dresses had, at some distance from the spectator, very much the appearance of a little troop of horse-soldiers. —An Icelandic lady's saddle is totally unlike an English one, being furnished with a semi-circular elevated back, like that which is attached to some of our old-fashioned chairs; so that a lady sits as much at her ease as the travellers in the passage of *Quindiu,* in the *Cordillera of the Andes,* who are described by M. Humboldt, as inhumanly riding in chairs strapped on the backs of their fellow men. Sitting sideways, therefore, the Icelandic women lean their backs against this support, and place their feet upon a small board, which is affixed to the saddle at a proper height by means of two straps. The back of the saddle is often richly ornamented with brass, carved or embossed into various figures: the girths, also, are furnished with rich silver buckles and with plates of the same metal, and the saddle-cloth frequently affords a beautiful specimen of the abilities of the owner at working in tambour.

it, and hear an Icelandic sermon. Accompanied by Jacob and my guide, I crossed a swamp which lay between us and the church; but, previously to entering it, we called upon an old lady, a rich farmer, who lives in the immediate vicinity, and whose hospitality is celebrated by Sir John Stanley. She was eighty-five years of age, and still enjoyed good health, though her faculties were much impaired, so that she scarcely recollected the visit of my countryman. A young man, however, whom she had adopted as her son, remembered him well. Her house, at this time, scarcely deserves the praises which Sir John has given it; for it was as dirty as any I had yet entered, and the closeness of the bed-room, into which we were ushered, was far from pleasant, and, I should suppose, equally far from wholesome. Yet in these confined rooms, where the external air is scarcely admitted, do the natives spend their time during the long winters, except, indeed, such of it as is necessarily employed in looking after their cattle; and here, too, by excluding the air, and by means of thick walls and a roof of turf, they are enabled to live without a fire in their sitting-room throughout the year. I heard the riches of the inhabitants of Haukardal much talked of: they consisted of ten cows, five rams, and about an hundred sheep; a property far from contemptible in this island, though scarcely more than equal to what Horace called upon his luxurious patron to offer at a single sacrifice on the safe return of Augustus, when, promising to sacrifice a calf for himself, he says to Mæceneas, "Te decem tauri totidemque vaccæ". An Icelandic church-yard is often in part enclosed by a rude wall of stone or turf, and the area, excepting only as much as is occupied by the building, is thinly sprinkled over with elevated banks of the green sod, which, alone, serve to mark the burial places of the natives, for whom no unlettered poet writes, or more unlettered sculptor carves, their names and years upon the monumental stone. This spot, previously to the arrival of the minister, on a sabbath, affords a most interesting spectacle. Numerous parties of men, women, and children, who had come on horseback, and in their best apparel, were continually saluting each other; and any person, that had been absent from the

the place of worship for a more than usual length of time, either through illness or any other cause, was kissed by the whole congregation. As they were little accustomed to see strangers, they all flocked around us, presenting us with milk and cream from the neighboring farm, and asking us an hundred questions. Many were surprised at our having come so far for the sake of seeing the Geysers, which they are accustomed to look at with the utmost indifference. The dress of the female children was like that of their parents, and some of them had even an equal number of silver ornaments: most of them wore the faldur, but some of the younger ones had, instead of it, small caps of black velvet or cloth, which fit close to the head and are tied under the chin, ornamented with gold lace, and frequently terminated by a silver gilt knob. Caps like this used formerly to be much more generally worn by the children than they are at present; and it is to be remarked, that not only the cap, but the faldur, also, when the wearer is on a journey, is carefully wrapped round with two or more chequered silk handkerchiefs, being preserved with the greatest nicety, as constituting a part of their dress, of which the Icelanders are particularly proud. Before the commencement of service, the priest read prayers to a woman after child-birth, who was sitting on a low stool at the church door; and this short ceremony was concluded by his laying his hands on her as she knelt. During the whole time, the woman seemed very much affected, and some who were standing round were extremely attentive. The church (which like most others in the island, fronted the west) was similar to the one at Thingevalle, but more commodious, in having benches instead of chests to sit upon. At the time I entered, the priest was at the altar, dressed in a long black gown of wadmal, buttoned from top to bottom in the front, black worsted stockings, and seal-skin shoes: his hair was hanging down a great length, reaching to his shoulders*. The women and young children alone sit in the body of the church, the men

* This is the case with all the natives, who consider it sinful to shorten the hair which God has caused to grow. It is for the same reason, I suppose, that a few, probably of the more orthodox, neglect to shave, and rather put up with the inconvenience of a long beard.

being ranged round the altar, near which, also, was reserved a place for Jacob and for me. It is these latter only that sing, if that monotonous and inharmonious noise which I heard on entering may be called singing, where every one strained his throat to the utmost, and gave out at the same time a most powerful effluvium of tobacco juice, which, mixing with the natural fish-like smell of the natives, rendered my stay among them in such a confined place by no means agreeable. As soon as the singing had ceased, one of the congregation put upon the priest a white surplice of unbleached linen, and over that a robe, on which was coarsely embroidered a large figure of a cross. He then chanted some prayers from a book, which, with more singing, performed sometimes sitting and sometimes standing, lasted about half an hour. Upon the altar lay a large snuff-box, a cup, and plate of silver, with a bottle of white wine, and a box of red wafers, not at all differing from such as are used in other countries to seal letters with. Of the first-mentioned of these articles the priest made frequent use: with the rest he prepared, during the time of singing, for performing the ceremony of the sacrament. He then ascended the pulpit, and, after repeating a few more prayers, delivered, in rather a quick but impressive manner, a sermon of about half an hour's length, after which the sacrament * was administered to the congregation, kneeling at the altar; to the men first, and then to the women; the priest putting a wafer and some wine into the mouth of every individual, and repeating at the same time a short prayer. This ceremony terminated the service, except the blessing and salutation, which were bestowed separately upon every one of the congregation, as well male as female. As soon as the whole was concluded, the priest spoke to us,

* The robes of the priest, and the wafers administered at the receiving of the sacrament, are institutions nearly related to those of the Romish church, and, together with the doctrine of consubstantiation, are maintained by the followers of Luther, whose religion is the established, and, indeed, the only one, of Iceland. The serious attention manifested by the simple natives during their devotions would have done credit to christians of any persuasion, or of manners however refined.

apparently much pleased at seeing strangers in his church; and, on hearing that we were about to set off for Skalholt in our way to Hecla, he begged that we would call at his house, which lay in the road, and would permit him to accompany us to that place, which we thankfully promised to do. Between the church of Haukardal and the hill Laugerfell, the morass abounded with the beautiful little *Ranunculus lapponicus* in full flower, a plant rare even in Lapland, the country whence it takes its name; while in the drier parts grew *Carex Bellardi*, forming a considerable portion of the herbage. At the northern foot of Laugerfell the minute *Konigia islandica* was in great profusion and perfection, as was also *Eriophorum capitatum* of Schrader, a species lately discovered in Sweden. On my return to the tents, I found both the Geyser and the New Geyser in pretty nearly the same state of fullness as when I left them, and they continued so till about eight o'clock, when there was an extremely fine eruption of the former. The day had been clear but cold, with the thermometer at 41°.

Monday, July 17. During the whole of the preceding night, both the Geyser and the New Geyser had remained perfectly quiet, but at four this morning we were gratified in seeing another eruption of the latter, equally magnificent as that of yesterday, though it did not last in all above an hour, and ceased spouting more abruptly than the former one: in every other respect the appearance was the same, and a second sight of this phænomenon did not at all detract from its impressive splendor. During my stay among the hot-springs, besides collecting a number of fine and beautiful specimens of the siliceous incrustation and other mineralogical productions, I filled from the Geyser and the New Geyser half a dozen bottles with water, none of which afterwards effervesced *, or was subject to any change,

* Unlike the water taken from the largest well of the springs near Reykum, in the parish of Huusevig, of which Horrebow relates, page 22, that, " if it is poured into bottles it will still continue to boil up twice or thrice, and at the same time with the water in the well. Thus long will the effervescence continue after the water is taken out of the well; but, this being over, it soon quite subsides and grows cold. If the bottles are corked up the moment they are filled, so soon as the

but continued altogether quiet, after having been corked, and tied round the mouth

water rises in the well they burst into pieces: this experiment has been proved on many score bottles, to try the effects of the water."—I fear there can be but little more credit attached to this story, than to the one which the same author mentions in a page or two following of his work, but which he has endeavored very candidly and justly to disprove. I allude to the account of a particular species of bird, which is affirmed, say Povelsen and Olafsen, by persons worthy of faith, to be found in the hot-springs, and not only to swim upon the surface of, but also to dive into, the boiling water. This ornithological rarity is reported to be of the form and size of a duck, of a deep brown color all over the body, except that there is a very conspicuous white ring round the eye. At the approach of a human being it dives and remains under water a long time, and sometimes, continue our Icelandic historians, it never comes up again! Much more curious matter, relative to this creature, may be found in the *Voyage en Islande, tom.* v. *p.* 88—89, whence I will extract a few lines, in the words of the author, or rather, of the French editor, for the amusement of such as may not possess the work. "Nous n'osons pas démentir tant d'assertions et des attestations de personnes dignes de foi; mais regarder ces oiseaux comme naturels, á combien de contradictions ne nous mettons nous pas en bute? Leur plumage, leur bec et leurs jambes défendus par une peau calleuse, pourraient, si l'on veut, supporter l'eau bouillante en nageant, mais en plongeant, que deviendraient leurs

with fresh sheep-skin *. Having now completed a number of sketches of the most interesting features of this remarkable spot, particularly of the two Geysers, and having concluded such notes and observations as I was able to make during my stay there, I found myself obliged to proceed on my journey. As I had learned that it was impossible to reach Hecla without first going

yeux?" After starting other probable objections to the power of diving in the hot waters, possessed by these birds, they conclude their remarks by saying, "nous croyons donc que si ces oiseaux existent, ce doit être des amphibies; en ce cas, ce serait une grande et intéressante nouveauté pour les naturalistes."

* These I had earnestly reckoned upon bringing to England for my chemical friends, and it was therefore a mortifying circumstance to me to find when, on setting out upon my second excursion, I was asked by our steward if he should fill again with water the bottles which I had brought in that state from my last journey, that he had emptied out every drop of what I had with so much pains secured, as I supposed, for a long voyage. I, however, dispatched some persons, with proper directions to the Geysers for another and a larger supply, and they actually brought back two horse loads, which shared a still worse fate on board the Margaret and Anne.

to Skalholt, at nine this morning our horses were loaded, and we started for that place. Frequently, as we went along, did I look back to catch a last glimpse of the scene which had afforded me a higher gratification than any thing I ever beheld in my life, and a pleasure which was only diminished by the want of an agreeable companion who could have been a partaker of the enjoyment with me: so true is the observation of the admirable French writer,—that every thing in this world, even solitude itself, loses half its charms, if we cannot have somebody to whom to tell how charming it is. At about twelve o'clock at noon, we arrived at the house of a kind peasant, whom we had seen at church the day before, and who, knowing we were this morning to pass his cottage, had stationed himself at the door with all his family, in their best dresses, to invite us to enter and take some refreshment. We were shewn into the fish-house, where cushions were placed for our accommodation upon one of the chests that hold their clothes, and some excellent new milk was set before us. From my host I first learned the difficulty, or, as he said, the impossibility of reaching Hecla after the late heavy rains, which would, in all likelihood, preclude any possibility of access to the mountain, or, even if the intermediate obstacles could be removed, and the base of the hill attained, would, most certainly, render climbing impracticable, by reason of the torrents of water rushing down on every side. I did not give much ear to this piece of information, though it was echoed by my Reikevig guide, who now began to show evident symptoms of fear at the prospect of visiting Hecla, and I determined, at all events, to proceed to Skalholt, as the only place where I should have a chance of obtaining more certain tidings, and guides to accompany me. Near this house I met with an itinerant beggar, of which there are many in Iceland; some of whom adopt this mode of life through idleness, and others through actual inability to do any sort of labor that might support them. The scanty supply of food which they necessarily procure by such means, in a country where even the most industrious are often reduced to a state bordering on starvation, renders these poor wretches real objects of pity and deserving of relief from travellers. I was surprised and mortified to find that this wretched being, who could scarcely crawl along, but who kept company with us some way on one of our relay horses, was not able to eat a morsel of the ship-bread and meat which I gave him; so accustomed had he been to a milk and fish diet, and such a stranger was he to a kind of food essentially different both in flavor and hardness. Our way lay over a great part of the same morass that we had previously crossed in a contrary direction after leaving the river Brueraa, of which stream we again came in sight during the course of this day's journey. We went only a little more circuitous route to see the hot-spring of Reykum, which I before mentioned, as visible to me at a distance. It is, indeed, in its present state, but little worthy of notice to any one, who had witnessed the finer eruptions of the Geysers: its water rises from a rugged aperture not more than two feet in diameter, and is thrown up to a height scarcely exceeding six or seven feet, but the spray is cast to a considerably greater distance; the jets are frequently repeated; and the eruptions are attended with a loud and rumbling noise, owing to the vast quantity of air which rises at the same time with the water. Some stepping stones in the river lead to a more quiet, but still hot, part of the channel, formed by this spring, and afford a convenient station for the people of the adjacent house, when they cook their provisions or wash their clothes. The inhabitants of this solitary dwelling, which is at a distance of about thirty yards from the fountain, assured me, that in the winter, in very clear and frosty weather, the height of the spout is sometimes so great, that, if the wind lies in a favorable direction for the purpose, the heated water and steam are driven into their house, to such a degree as to compel them to seek a temporary residence elsewhere. Inconsiderable, however, as I found this spring of Reykum, it, nevertheless, with its accompaniments, forms an object both beautiful and interesting, amid the dreariness of the surrounding scenery. The grass, growing near its margin, was longer and more luxuriant than almost any I had seen in the country, and some little rocky islands in the stream, a few yards below the crater, were clothed with a rich bed of *Sphagnum latifolium,* intermixed with *Hydrocotyle vulgaris,* and the elegant little *Epilobium alpinum,* then in full blossom. Our course had hitherto been westerly, but we now turned our faces to the south, and looked towards Skalholt, pursuing a tolerably good track, which led us through a less boggy soil to the house of the priest whom we had met at Haukardal, and whom we now found busily engaged in cutting peat * from a neighboring morass for his winter fuel, dressed in clothes made of undied worsted, with a long blue cap upon his head. The church, hard by, however, which contained his wardrobe, afforded this worthy man a suit of black wadmal, in which he attired himself to accompany us to Skalholt. It required some caution to wade through the morass which lay between us and that place, but the immediate entrance to the small cluster of houses that com-

* The instrument used for this purpose is called *Torf-Liaar,* and is well figured in the Atlas of the *Voyage en Islande, tab.* 8. *f.* 3. In shape it is not much unlike an instrument used in this country for cutting hay on the stack, and it is employed in the same way.

posed this village, which was but a few years ago the residence of the bishops, and the capital of Iceland, was, if possible, still worse, being an extremely wet and boggy soil, interspersed with large pieces of rock. One good turf house, and three or four smaller ones are, besides the church, all that now remains of the town. The adjacent country is by no means pleasant, though grass is tolerably abundant. Immediately surrounding Skalholt we remarked the ground formed into a number of little hills, among which was to be seen here and there the steam arising from some hot-springs, and on the opposite shores of the river Hvitaa, which is here of considerable width, is situated a small and rather grassy mountain. In the south-east, over a low range of hills, Hecla reared its head full in our view, covered with snow, more than half way down from the summit. We had scarcely pitched our tents, when a handsome young widow, of the name of Joneson, richly dressed in the Icelandic fashion, came down and invited us to her house, where she set before us some *Ren*, or rye-pottage, in a turenne, and a basin of cream and sugar. It was one of the best Icelandic houses I had ever entered, and was, moreover, in every part remarkable for its extreme cleanliness, in which respect our hostess herself was no less conspicuous. The rooms were wainscotted and painted with blue and red, and there was a good library, belonging, however, to the school of Bessestedr, the lector of which place, who was brother to Madame Joneson, frequently consulted it. The collection contained many of the classics, but consisted chiefly of Icelandic books and manuscripts, relating to the political and ecclesiastical history of the country, mixed with extracts from such works as are most scarce in the island; among which I noticed several pages copied from the Linnæan *Amœnitates Academicæ.* The farm, belonging to this house, was reckoned a considerable one, and had several buildings appropriated to the use of cattle; but of these, the floors are never covered with any sort of litter, so that the poor animals must have but a sorry bed on the bare rock. From the exceeding filthiness of the place, it seemed as if a dung-hill, near the outside of the building, was but seldom replenished. At Skalholt, for the first time, I saw people cutting hay; which they do by means of a scythe * with a straight stem, about six feet long, from which project, at right angles, two handles, and, as the ground producing their crop of hay is broken into innumerable hillocks, they find it advantageous to use a blade of not more than two feet in length, with which they perform the operation more in the manner of chopping up the grass than mowing it. In the evening, I met with a truly wretched object, a woman who was afflicted with the malady called among the Icelanders *Likthrau;* a species of leprosy, or more properly, according to Von Troil, elephantiasis. Her face was so corroded by the disease, that it presented the most disgusting spectacle I ever saw in my life, and her legs and hands were swollen to an enormous

* A scythe, in every respect resembling this, is used by the natives of East Bothland. A description and very accurate figure of one are given in the second volume of the *Lachesis Lapponica,* where the author remarks, that this instrument possesses the advantage of enabling the mower to move forward in nearly an upright posture.

size, these latter being, also, covered with a thick and almost white skin, lying in great wrinkles; yet she still complained of no particular pain, and seemed to walk with tolerable ease. This terrible complaint is well known to be hereditary, but it nevertheless frequently happens, that the children of those affected are, for many generations, quite free from it; an instance of which presented itself in the son of this very woman, who was constantly with her, and yet shewed not the least symptom of the malady; but, on the contrary, was one of the most healthy and beautiful children which this country had offered to my view. Neither, indeed, does it appear to me to be infectious, any more than another cutaneous disorder already mentioned as common in the island. It is said to have existed in Iceland ever since the first colonization of the country, and is supposed by many to have been brought over at that time from Norway, where, according to some accounts, it may be traced to a period of high antiquity. Its prevalence and virulence are, probably, in a great degree, ascribable to the use of woollen clothes*, and to the mode of living and habits of the natives; for they take but little exercise, except in the fishing-season, when they are continually wet with salt water; and their food is peculiarly calculated to promote scorbutic affections, consisting, at the time of fishing, almost entirely of fresh fish, and at other times of dried fish, in both cases generally unaccompanied with vegetables. The inhabitants of the canton of Bardestrand, and those who live near Patrix-fiord, are said to be in the habit of making use of antiscorbutic vegetables, and to be, consequently, more free from the disease. The plants that I met with about Skalholt, were such as I had elsewhere seen, excepting only one or two grasses, which appeared new to me. *Ranunculus lapponicus* was here very abundant, as was the *Konigia*, and a new species of *Carex*, which I had before met with near Reikevig. On the walls of the houses grew *Draba contorta* and *Tortula subulata*: *Angelica archangelica*, too,

* The elephantiasis used to be equally prevalent in Great Britain, previous to the introduction and adoption of linen, instead of the woollen clothes then universally worn.

was not less plentiful here than in very many other parts of the country; but, although certainly employed as an esculent plant, both fresh, and after having been kept for some time buried in dry earth, and eaten with fish or butter, yet it is by no means so universally adopted, as is the case among the natives of Lapland *. This whole day was fine and warm, except that towards

* This plant is not only an article of luxury among the Laplanders, but is also used by them as a medicine to cure the spasms, arising from cholic, on both which subjects Linnæus expresses himself so beautifully, in his own peculiar language, in the *Flora Lapponica,* that I make no apology for the length of the following extracts:—" Morbo laborant sæpius Lappones sylvatici vehementissimo, *Ullem* vel *Hotme* dicto, qui species Colicæ est, et ad Colicam spasmodicam Scheuchzeri proxime accedit; corripiuntur enim interanea circa regionem umbilicalem spasmis dirissimis, qui extenduntur ad pubem usque, paroxysmis parturientium sane vehementioribus, ita ut miser Lappo, vermis instar, repat per terram et urinam sæpe sanguinolentam reddat, licet calculi nulla omnino sit suspicio apud hanc gentem a calculo et podraga privilegiis naturæ defensam; post aliquot horarum, quandoque diei, spatium, resolvitur, ptyalismo ingenti per quadrantem horæ durante. Dicunt ipsi, quod hic morbus in Alpibus eos non adgrediatur, sed tantummodo dum in sylvis per æstatem degunt, hausta scilicet ibidem aqua

the evening a thick misty rain came on. At nine in the morning, the thermometer was at 60°.

semiputrida, vi radiorum solarium calefacta, vel forte vermiculis scatente. In hoc morbo variis utuntur medicamentis, et omnibus quidem fortissimis, ut vehementem morbum æque vehementi oppugnent alexiterio, quale est radix Angelicæ, cineres aut oleum Tabaci, Castoreum liquidum, &c.—Caules Angelicæ hujus sunt Lapponum deliciæ et fructus æstivi, quibus benigna natura eos donavit, dura nimis et immisericordi existente Pomona, quæ Lapponum terram nunquam intravit. Caulis hic, antequam umbella absolute explicata est (nam circa florescentiam lignosus evadit), abscinditur prope terram, folia avelluntur et cortex ad basin caulis cultro dentibus vel unguibus solvitur, detrahiturque a basi ad apicem, cannabis instar, remanente interiori caulis parte nuda nivea concava et pulposa, quæ instar Rapæ vel Pomi cruda editur et quidem summo cum adpetitu, deficiente gratiori in hisce oris vegetabili. Cum pueri vel puellæ mense Julio cum rangiferis suis per Alpes errantes in pascuis illosque circa vespertinum vel matutinum tempus ad casam, ut mulgantur, reduces comitantur, detruncatis caulibus totum sinum impletum reportant, quos in familia sua distribuunt, et summa aviditate devorant. Gratus hic Lapponibus cibus nec nobis displicebat, leviter enim amarus et simul aromaticus est, immo et gustui et ventriculo arridebat, adsumptis scilicet tamdiu diluentibus ac emollientibus, cibo non salito, carnibus et piscibus sale nullo maceratis, lacte pingui rangiferino,

Tuesday, July 18. Coffee was early prepared for us by Madame Joneson this morning, and was succeeded by a glass of rum, previously to our taking our breakfast, which consisted of a large dish of boiled salmon, eaten with butter and vinegar, and, after it, a mess of mutton, boiled to rags, mixed with melted butter, and seasoned with a sweet sauce of oatmeal and sugar. During this repast, the persons, who were sent for the preceding evening to be my guides to Hecla, arrived with the unwelcome intelligence, that, in the present state of the weather and morasses, they neither could nor would undertake to conduct me to that place. The rivers, too, were so swollen, that those, which at other times were said to be deep, were not now to be crossed without extreme danger. My Reikevig guide, also, declared he would not proceed with me, but await my return at Skalholt. It was in vain con-

haustaque pura puta aqua; tum, inquam, optime conveniebat, sed nescio num in hortis nostris magis amara sit et acris, vel an gustus nobis in Lapponia fuerit alius, quam extra eam; extra Lapponiam enim nunquam arrisit: forte fercula persica persicum requirunt adpetitum." *Fl. Lapp.* p. 73.

tending with the obstinacy and superstitious timidity of these men; for, though, owing to the excessive wetness of the season, there would, undoubtedly, have been some difficulty in wading through the morasses, yet their apprehensions principally arose from the necessity there would have been for them to climb a volcanic mountain, which many of them believe * to be the abode of the damned, and which all the lower class

* This opinion is well known to have existed of old in heathen superstition; following which the classical poets make Ætna the prison of the giants: Gaspar Peucer, as quoted by Arngrimus Jonas, states the matter, respecting Hecla, very circumstantially: "Est in Islandiâ, inquit, mons Hecla, qui immanis barathri, vel inferni potiùs profunditate terribilis ejulantium miserabili et lamentabili ploratu personat, ut voces plorantium circumquaque ad intervallum magni miliaris audiantur. Circumvolitant hunc corvorum et vulturum nigerrima agmina, quæ nidulari ibidem ab incolis existimantur. Vulgus incolarum descensum esse per voraginem illam ad inferos persuasum habet. Inde cum prælia committuntur alibi in quâcunque parte orbis terrarum, aut cædes fiunt cruentæ, commoveri horrendos circumcirca tumultus, et excitari clamores atque ejulatus ingentes longâ experientiâ didicerunt." *Hackluyt's Collection of Voyages*, edit. 1810, vol. ii. p. 590.—Not very dissimilar is the vulgar belief among the Japanese, except that

of people regard with the greatest horror. Although I had been informed by Icelanders of respectability, who had visited this mountain, that I should see nothing remarkable upon it, but what I had seen elsewhere, still I felt a great mortification at the refusal of the guides to accompany me; because, next

they, instead of imprisoning their damned in the volcanoes, consign them to the boiling fountains; upon which subject Kæmpfer has the following curious remarks :—" The monks of this place (Simabara) have given peculiar names to each of the hot-springs, arising in the neighborhood, borrowed from their quality, from the nature of the froth at top, or the sediment at bottom, and from the noise they make as they come out of the ground; and they have assigned them as purgatories for several sorts of tradesmen and handicraftsmen, whose professions seem to bear some relation to any of the qualities above mentioned. Thus, for instance, they lodge the deceitful beer and sackibrewers at the bottom of a deep muddy spring; the cooks and pastrycooks in another, which is remarkable for its white froth; wranglers and quarrelsome persons in another, which rushes out of the ground with a frightful murmuring noise; and so on. After this manner, imposing upon the blind and superstitious vulgar, they squeeze money out of them, making them believe that by their prayers and intercession they may be delivered from their places of torment after death."—*History of Japan*, vol. i. p. 106.

to visiting the hot-springs, the opportunity of climbing Hecla was my grand object in Iceland. At first, I thought of waiting a few days for better weather, but the continuance of the rain, and the little prospect there was of its clearing up induced me, before the evening, to determine upon departing for Reikevig on the morrow; especially as the fortnight, the time allowed me previously to the sailing of the Margaret and Anne, was within three or four days of its expiration. However, I left it with somewhat the less regret, from hoping it would be in my power to revisit the country at a future time, under more fortunate auspices. I have before mentioned that the bishop's see had been removed from Skalholt to Reikevig: at the same time the cathedral, also, was pulled down, and a new and very neat wooden church erected in its stead. Our fair hostess accompanied us to this building, which, however, contains none of those reliques of antiquity * that

* These were, at the time when Olafsen and Povelsen wrote their history (about 1760), two ancient altar-pieces, and a bishop's staff (bâton d'Evêque), of which the upper part was brass, richly gilt. There was, like-

the cathedral was said formerly to possess, unless, indeed, an altar-cloth, with some robes, and a mitre richly worked in gold, but now very much tarnished, may be con-

wise, to be seen the coffin of St. Thorlak, who was made bishop of Iceland, in 1178, and died in 1193. His *Saga* is said to be full of miracles, and he found worshippers, according to Von Troil, not only in Iceland, but also in Denmark, Norway, England, Scotland, the Orkney Isles, and Greenland, and even had a church dedicated to him at Constantinople. On the thirteenth of August, 1198, his bones were dug up and deposited in a coffin, plated with gold and silver, and it was resolved that this day, as well as that on which he was elected bishop, and that on which he died, should be annually celebrated. Gysserus Einarsson, who was made bishop in 1540, and was a violent enemy to popery, caused the ornaments to be broken off, and the coffin covered with copper gilt: in such state it was exhibited in the cathedral at the time Sir Joseph Banks was there (1772). The relique that was shewn for a portion of his skull was ascertained to be only a piece of a large cocoa-nut-shell!—While preparing this part of my little work for the press, I have been enabled, through the kindness of Sir Joseph Banks, to have before me, amongst many other drawings made by his artists, two, which represent views in different directions of the cathedral of Skalholt: from these it appears to have been built entirely of boards, in the form of a cross, and, but for a little wooden spire, would have been so like an English barn, that I do not know any thing with which I can so well compare it.

sidered as laying claim to be so regarded: unfortunately my memory, at this time, will not enable me to recollect what I was informed concerning them. The pulpit in the church is extremely well made, and some small, but not ill executed, figures, are painted upon it. A very tolerable Danish painting, also, of the late bishop of the place, who had, if I mistake not, married a sister of Madame Joneson, is hanging up against the wall; and, underneath the floor, which affords a protection to it from injury, and of which a part lifts up, like a trap-door, to exhibit it, is laid a handsome tablet, richly inscribed in gilt letters, in commemo-

The numerous small buildings that were then situated close by the cathedral, and formed the town, were occupied, as Sir Joseph Banks informs me, entirely by the bishop's dependants and twenty-eight boys who were at the school, and were maintained at the expence of the King of Denmark. Among the whole cluster, I can now only recognise the house at present occupied by Madame Joneson; so much is the place altered within these forty years.—Sir Joseph also possesses the drawing of an ancient weapon, seven feet long, which he saw in the cathedral of Skalholt, in shape much like a halberd, and said to have belonged to a famous hero named Skarphedin, who died in the year 1004.

ration of his virtues and learning. The cathedral of Skalholt is reported to have been a noble structure, and perhaps really was so for Iceland, where the magnificence of buildings is not to be estimated according to our southern ideas; but the foundation, which still remains, and may be traced extending some paces beyond that of the present building, does not appear to admit of its having been what we should call in England a large or even a moderately-sized edifice. It was in the year 1057 that a bishoprick was established in Iceland at this place, only eighty-three years after the introduction of christianity, till which period the natives were worshippers of idols. The bishop that first filled the see was Isleif, the son of one Gissur, who, together with a person of the name of Hialte Skeggesen *, preached the doctrine of christianity with

* Of these persons Povelsen and Olafsen relate the following anecdote from the *Khristni-Saga.*—" It was Oluf Tryggveson, King of Norway, who, after having been at much pains to induce the Icelanders to embrace the new religion, sent them these two men to complete the work; but their proceedings were near failing of the purposed end; for the volcanic eruption then took

so much success, that, at a general convocation of the people of the island, held at the Althing in the year 1000, it was agreed, that idolatry should be abolished, and the religion of our Blessed Saviour adopted in its stead. The many kind attentions, and the truly hospitable entertain-

place which produced the lava called *Thurraarrhraun*, and just at the time when they were preaching to their countrymen, some messengers arrived with the grievous intelligence †: whence the pagans were led to believe that they saw in this eruption a proof of the anger of their gods, at the blasphemous discourses of the partisans of christianity. It was not a little fortunate then, that at a moment, as critical as it was decisive, one of the pagans named Snorro Godi, a priest (who, perhaps, had conceived a good opinion of the new faith), succeeded in calming them, by putting to them the following question, no less laconic than ingenious: 'um hvat reiddnust gudin tha er her brann raunit er na floendum ver a?' *What, then, was the cause of the anger of the gods, when they burned the rock on which we are now standing?*—for all who were present knew that this had happened before the country was inhabited."

† "Ecce autem vir cursu anhelus: ignem subterraneum in Olfus erupisse, et jam villæ Thoroddi pontificis imminere nunciat. Tum ethnici: non mirum, si ejusmodi sermonibus excandescerent Dii, vociferantur. At Snorrius pontifex: '*quid igitur excanduerunt Dii, cum scopulus cui nunc insistimus conflagravit?*' " *Khristni-Saga*, cap. ii. p. 88—90.

ment which I had received from Madame Joneson, made me feel anxious to make her in return some little acknowledgement, and I was vexed, on examination of my stock, to find it so much reduced as to render it not a little difficult what to fix upon that might be acceptable. My tea and coffee were already expended; nor could I think of any thing to offer her but a shirt, a few cravats, and a pocket handkerchief. I felt how unworthy such trifles were of her acceptance, as a reward for so much hospitality, and I was therefore the more pleased to find them received with evident marks of gratification. Her happiness was manifested by a friendly salute, and by the eagerness with which she unfolded and surveyed the different articles. She was greatly puzzled, however, to ascertain the use of the frill of the shirt, and led me into no less perplexity by consulting me on the best mode of converting it into an article of apparel that might be serviceable to herself. I was much struck with this incident, as singularly characteristic of the simplicity of manners even of the higher classes of the inhabitants, and, trifling as it may appear in itself, I therefore record the anecdote in my journal. A rainy afternoon made me come to a determination to turn our backs at once on Hecla *, and return without delay to Reikevig, in pursuance of which, about six o'clock in the evening, having struck our tents and procured guides, we took leave of our kind hostess at Skalholt, and set out upon our journey, proposing to travel on, keeping along the south side of Apn-vatn, till we reached Thingevalle. The first part of our route was truly execrable, lying over rocky hills, whose surfaces were every where strewed with loose angular pieces of stone. A steep descent brought us to the banks of a deep and wide river, where we found a miserable conveyance for ourselves and luggage in a boat which had been formed out of half a larger one, and was so leaky as to require continual baling till we reached the opposite shore. Our horses were obliged to swim, which they did with great dexterity, keeping only their noses above the water, though carried by the rapidity of the current a considerable

* See Appendix C. for an account of this mountain.

way down the stream. An extensive rocky moor succeeded, interspersed with disagreeable bog and numerous rivulets, and presenting nothing interesting to the traveller, till, about ten o'clock, our wearied eyes were relieved by the view of Apn-vatn, and of a lofty column of steam from the boiling fountain of Reykum. As we ascended the hills on the west side of Apn-vatn, the rain changed to a thick mist, accompanied by a degree of cold, which I should scarcely have thought could have been experienced south of the arctic circle in the month of July. A flannel under-dress and two great coats, in addition to my usual quantity of apparel, were not sufficient to keep me warm, and I frequently found it necessary to alight from my horse, preferring the fatigue of walking under such a weight of clothes, to the excessive cold experienced during more moderate exercise. About midnight it became apparent from the broken surface of the ground, and the holes which here and there presented themselves, that we were approaching a continuation of the extraordinary country that extends in an easterly direction from Thingevalle, while the duskiness that pervades the atmosphere in the night at this season of the year, together with the fog which now confined our view to within a few yards around us, but which at the same time increased the apparent size of the objects, added to the gloominess and horror of the scenery. We travelled continually among the great masses of rock that lie strewed in the wildest possible disorder about the chasms which they once served to fill up, and frequently, as we went on, were we deceived by the imaginary sight of houses in this solitude, which, on a nearer approach, proved to be only huge rocks, torn from their natural situation by the shock of an earthquake, or some terrible convulsion of nature. However naturally the mind of man shrinks from solitude, and rejoices amidst the dreariness of an Icelandic waste to see the faces and to hear the voices of human beings, yet still in a country like this, where the track, whenever it appears, affords room for only a single horse, the sense of danger overpowers the gratification, and it is therefore fortunate that travellers are seldom met with, except at this season, when the natives are returning from the mart at Reikevig, or from some fishing station on the coast, bringing with them their supply of fish, and of other articles necessary for their subsistence or convenience. Such a party, loaded with planks for building, we here heard at some distance before us, urging their fatigued beasts to quicken their pace; and their toil was increased by their being obliged to pass us in a place where the excessive inequality of the surface would effectually have stopped the progress of any but Icelandic horses. Soon after this we approached a rocky mountain, at the south-east end of Thingevalle-vatn, and, shortly after came to the margin of the lake itself, where, by keeping as near the shore, as the nature of the country would allow, we escaped the worst part of the chasms, which we had some days before experienced so much difficulty in crossing; and we enjoyed, as the mist dispersed, about two or three o'clock on the following

Wednesday, July 19.

morning, a magnificent view of Thingevalle-vatn, with its two black islands; whilst we ourselves were riding along the banks amidst a small copse of diminutive birch, intermixed with alpine willows, and varied with the bright blue of the flowers of *Geranium sylvaticum*, which grew here in considerable quantity. For a few minutes we stopped to bait our horses in this verdant spot, and then continuing our way over a track of country that I have already attempted to describe on my road to the Geysers, at about five o'clock we came to the house of the priest of Thingevalle. Unwilling, however, to disturb the family at so early an hour, we crossed the Oxeraa, and once more entered my favorite spot of Almannegiaa; here proposing, if the weather would allow of it, to spend two or three days. No sooner was our little encampment completed, than I clambered over some loose pieces of rock, which, crossing the chasm, formed a slight barrier; and hence proceeded about a mile up the southern part, where I found that, on the west, the perpendicular face of the rock increased in height as I went along, while the opposite or eastern side was in many places not a quarter so high. Indeed, in every part of this chasm that I examined, the western side was the most lofty, and was quite perpendicular, but the eastern constantly very considerably less in its elevation and leaning outward, so that a section of the chasm would represent the annexed figure.

Among the rocks grew, rather plentifully, ***Polypodium hyperboreum,*** and a species of fern which appeared to me new, but of which I do not sufficiently recollect the characters to attempt a description of it. On climbing the eastern cliff, and descending on the grassy surface to the margin of the lake, I found, but sparingly, *Isoetes lacustris.* As nothing more remarkable invited me to proceed in a southern direction in the chasm, I turned to the north, and retraced my steps; when, on looking back, after walking a few hundred paces from my tent, I was amazingly struck with the terrific appearance of the entrance of the pass of Almannegiaa, the descent through which I have previously mentioned. Huge masses on the summit of the precipice scarcely appeared to be attached to the edge on which they stood, so that you would think the slightest breath of wind would hurl them into the plain below; while all around, in addition to these, the great fissures, the rocks projecting from the sides, and the scantiness of vegetation, formed a scene truly grand, but at the same time divested of every thing that might be called beautiful. Farther to the northward I met with several plants which I had not before seen in the country: among them were *Saxifraga cernua*, a new *Marchantia* in fructification, two or three *Hypna*, with which I was unacquainted, and *Fontinalis squamosa*, also, full of capsules. The noise which I now heard of the falling ot water convinced me I was arrived in the neighborhood of a cascade, of a portion of which I had previously caught a distant glimpse, sufficient to awaken my curiosity and make me feel anxious to approach it; to effect which it was necessary to cross one or two rapid torrents, when, turning round a projecting angle of the cliff, I had suddenly a full view of a very magnificent cataract, dashing its foaming waters with tremendous roar over the highest part of the precipice, whence they fell in an unbroken sheet upon the rocky base, composed of immense masses of most uncertain sizes, all rounded and rendered perfectly smooth by the force of the current, which, after crossing the chasm in an obliquely winding course, makes its boisterous way through a most romantic opening in the eastern cliff, and then soon unites with the more quiet stream of the Oxeraa *, at about half a mile from its confluence with Thingevalle-vatn.

* I have, on my first visit to Thingevalle, mentioned that it was the spot where the court of justice was held, and that near it was the place of execution for criminals. Since that was written, I am informed by Sir Joseph Banks of a peculiar punishment formerly inflicted upon women for the murder of their illegimate children. "They are drowned," says Sir Joseph, in his

At the distance of a few hundred yards from this cascade lay some pieces of rock, which had fallen from the cliff, in such a manner as to enable me, though not without considerable difficulty, to reach the summit, where I had an opportunity of seeing the stream which supplies the waterfall, as it rolled rapidly, a deep and wide mountain-torrent, through a nearly level bed of unproductive rock. The upper surface of the cliff, as far as I could see, both on this and the opposite side *, may clearly be perceived to

journal, "in a pool in the river, under a cascade; examples of which are very scarce, but one happened in the youth of the clergyman of Thingevalle, who was (in 1772) fifty years of age. The criminal was tied up in a sack which came over her head, and reached as far down as the middle of her legs; a rope was then fastened to her, and held by an executioner on the opposite bank: after standing an hour in that situation she was pulled into the water, and kept under with a pole till she was dead."

* From the summit of the eastern cliff there is, as I have before remarked, a sudden declivity into the great plain in which Thingevalle-vatn is situated, and not only the surface of this is curled, and bears the most striking marks of volcanic origin, but, as Sir Joseph Banks was informed, the bottom of the lake, also, exhibits the same appearance.—The fol-

have been in a melted and flowing state from its curled appearance, and in the face of the precipices the different currents of lava are very visible, of various thicknesses,

lowing remarks and sketch, from Sir Joseph Banks' journal, will assist in rendering more intelligible my description of Almannegiaa. "The highest cliff was ascertained by measurement to be one hundred and seven feet six inches, the opposite one thirty-six feet five inches, and the width of the chasm one hundred and five feet. The face of the precipice presented to our view, currents of lava, varying in thickness from ten inches to as many feet, each of them being distinguished from the other by its curled and porous surface. Some of them form arches, having run in hills: all of them, probably, proceeded from one eruption, though in different streams. The lesser height and oblique position of the eastern mass, and, indeed, the chasm itself, it may be conjectured, were caused by some under stratum having given way, and the consequent sinking of all above it, as the figure will better explain."

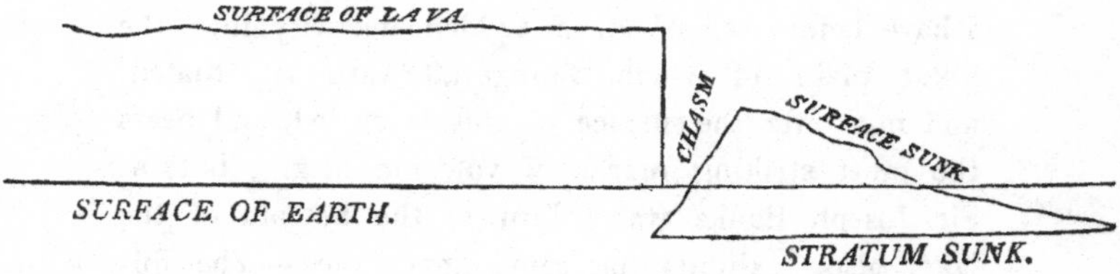

divided here and there by perpendicular fissures. A heavy rain now put a stop to my botanizing, which was, indeed, become an useless occupation, as all the specimens that I might have gathered would necessarily have been destroyed, and I therefore returned to the tents, whence Jacob and myself took our horses to call upon the priest of Thingevalle, for the purpose of making some acknowledgement for the kindness he had shewn us. Near his house I was much struck with the venerable figure of a native, who was employed in cutting the twigs of birch into small pieces, for burning into charcoal. His long beard and the singularity of his dress gave him very much the appearance of the Icelander represented in the ancient costume of his country, in the third plate of the Atlas of the *Voyage en Islande*. His jacket was ornamented with a coarse sort of lacing, and his little hemispherical cap, fitting close to his head, was precisely the same as the one there figured. This old Icelander served likewise as fisherman to the priest, and had just drawn from the lake a considerable quantity of the Thingevalle trout, which are, at this season, to be taken in the greatest abundance; yet, it nevertheless does not appear that any means are employed for the purpose of curing them for a winter stock, in which state they might afford nourishment to a number of poor people who reside in the neighborhood. Indeed, I do not recollect seeing throughout this extensive piece of water more than two or three boats engaged in the fishery, and the peasants who lived only a few miles distant from Thingevalle-vatn seemed scarcely to know of the existence of such a fish as the forelle. A vast heap of *Lycopodium alpinum*, lying near the priest's house, drew my attention, and, on inquiry, I found that it was used for the purpose of giving their wadmal a yellow dye *, which is done

* For giving the same tint to woollen cloths, according to Povelsen and Olafsen, the inhabitants of Borgafiord and its neighborhood make use of the *Lichen islandicus* in the following manner: they strew some of it upon the surface of the stuff intended to be dyed, to which it readily attaches itself, and they then roll the cloth upon a cylindrical piece of wood and boil it for six hours in an iron pot; which done, they take it out of the water, unrol it, and lay it in the air to dry:—the color thus acquired is a dark, but excellent, yellow. A deep brown dye is produced by boiling the cloth in

by merely boiling the cloth in water, with a quantity of the *Lycopodium*, and some leaves of *Vaccinium uliginosum*. The color, imparted by this process, to judge from some cloth shewn me, was a pale and pleasant,

water with a quantity of the leaves of the *Sortilyng* or *Arbutus Uva Ursi*, in the same way as practised with the *Lichen islandicus* ; and in case it is afterwards desirable to make this cloth black, some fat earth of that color, called *Sorta*, is collected, put into a vessel of water, and stirred about briskly, till it has acquired the consistency of paste; in which state, if suffered to stand a little time, the lower part stiffens into a thicker substance, and a liquid floats on the surface, which being poured off, what remains is daubed over the cloth whilst the leaves of the *Sortilyng* are still attached to it: the cloth then, having been rolled upon a cylinder, is boiled, together with the paste, for some hours; taken from the vessel; suffered to cool, and washed in fresh and cold water. Dr. Westring, in his admirable work upon the dying qualities of various Lichens, has given a figure of *Lichen islandicus* and specimens of four colors that may be extracted from it, by different processes ; a pale bright yellow, a rusty red, and two modifications of brown.—I am happy in the opportunity of recommending to the attention of my countrymen this beautiful and elaborate performance, a translation of which (from the Swedish language in which it is written) might possibly be of considerable service to some of our British manufacturers.

though not a brilliant, yellow. A similar dye is said, by Linnæus, to be produced in Lapland from another species of *Lycopodium*, the *L. complanatum*, but with this, instead of the *Vaccinium*, are used birch-leaves, gathered at midsummer.

Thursday, July 20. Owing to the continuance of the rain, it appeared to be useless to remain longer in Almannegiaa; therefore, after spending the day in making such sketches of the most remarkable parts of the scenery as the weather would allow, I resolved to depart myself with Jacob, at six o'clock in the evening, for Reikevig, leaving my guide, with orders to follow me as soon as the tents and luggage should be sufficiently dry. The margin of Thingevalle-vatn fortunately enabled us to find our way to Heiderbag; for, otherwise, we should in all probability have been lost for a time, owing to the excessive fogginess of the atmosphere, which would infallibly have prevented us from reaching the house of the pastor Egclosen, where it was necessary for us to procure a conductor for the following part of our journey. Indeed, as often as our leaving the shore and deviating from the track induced the necessity of Jacob's being separated from me, in order to recover it, it was only by shouting to, and answering one another, that we were enabled again to join company, so thick is an Icelandic fog, of the influence of which it is scarcely possible for an adequate idea to be conceived in England, except by those who have had the misfortune to be in the crowded streets of London in similar weather. At nine o'clock we arrived at the door of the worthy priest, whom we found seated in the fish-house, nursing his infant child, and at the same time employed in preparing his discourse for the following sabbath. A man, who was engaged stowing some fish and wool, in the same building, offered to accompany us on our way, and the priest immediately sent him in search of his horse, which was grazing on the morass. The rain and fog had by this time so increased, that we gladly availed ourselves of the shelter before us, and partook with thanks of such refreshment as our host was able to afford. After three hours passed in anxious expectation of the return of our guide, we at length began to fear lest some accident should have befallen him; for the animal could not have strayed far enough to detain him any great length of time; not only because the spot that would afford the poor beast nutriment was very circumscribed, but because it was fastened by its fore legs. The priest, however, did not partake of our fears, but was more inclined to think that the intensity of the mist had prevented the man from discovering the horse, a circumstance far from impossible, although he might be within a few yards of him; and, to convince us of the probability of his conjecture, he told us an anecdote of a person, whom he knew, being, during the continuance of an equally thick, but more durable, fog, for two whole days engaged in a similar search. The conjectures of the priest respecting our guide were indeed well-founded; for at twelve o'clock he returned with tidings that he had not been able to find the animal, and he therefore volunteered his services to conduct us on foot beyond the most intricate part of our route, an offer that we gladly availed ourselves of, as to have gone to the nearest neighbor to borrow a horse would have occupied full three hours. A glass of rum, with the flavor of which our guide seemed scarcely to be acquainted, and of the strength of which he had no idea till he had drunk it, had such an effect upon him, that he did not seem to need a horse to carry him faster, and he continued running for more than an hour without once stopping; except, indeed, when he was so unlucky as to strike his foot against a stone, and fall, in consequence of it, among the rocks. This circumstance frequently happened, and at every time he looked back and laughed, as if sensible of the cause of his stumbling; always taking care to tell us he was not hurt, and proceeding immediately with his previous speed. He several times forded rivers whose waters reached as high as his waist, and tried, by wading in different parts of the stream, to find the shallowest and least rocky places, so that we might be enabled to pass with the greatest ease and security. As often as we had to cross a morass, he went before us with a long pole and pointed out the unsound spots, which, however, without this precaution, the sagacity of Icelandic horses is almost sure of being able to discover; for, if they perceive, by a difference in the vegetation, a part which appears insecure to tread upon, they immediately put their noses to the ground, and, as if by the faculty of smell, seem to be sensible of the propriety or impropriety of proceeding. This instinct, indeed, is not peculiar to the horses of this country, for the shelties of Scotland appear to possess it nearly in as great a degree. After conducting us into a beaten track, at about three o'clock in the morning of the following day, our attentive guide left us, and with no diminution of speed set off on his return to Heiderbag, in order that he might reach the place in time to go through his whole day's work of hay-cutting. The mist now began to clear away, and I saw at but little distance before me the chasm at the foot of the mountain Skoul-a-fiel. I alighted from my horse and walked along a steep descent to the edge of the precipice, whence I looked directly down into an opening of the ground, which, at the same time that it appeared nearly as deep and quite as terrific as that of Almannegiaa, was more remarkable, from having

Friday, July 21.

in the centre, between the two precipices, a perpendicular column of rock, in height nearly equalling the place on which I stood, and surrounded, excepting a small portion, by the waters of a torrent that flowed with great rapidity along the bottom of the chasm. There was no way by which I could arrive at the stream without going a very circuitous route, and I therefore thought it better to hasten to Reikevig, and, if the time allowed me before the sailing of the vessel would permit, to return and bestow a day upon the investigation of this place and the neighboring mountain. On our nearer approach to Reikevig, we saw numerous parties of the natives with their tents and horses, giving an appearance of life and population, which alone could make the rest of our journey in the least interesting. Wheresoever a green spot presented itself, tents were pitched, and the horses suffered to graze, whilst the owners were reposing themselves after a journey which had been made during the night, according to the general custom of the Icelanders at this season of the year. These people were on the road either to or from Reikevig; in the former case conveying the produce of their flocks or wild animals; in the latter, bringing back articles of foreign manufacture, or, as is most usual, fish for their winter's supply. Among those returning from the mart, I recognized my young friend, the son of the priest of Thingevalle, who had been disposing of a cargo of butter and wool for his father. In passing by such a collection of Icelanders, amounting to many hundreds (a number which, for the the space of a month in July or August, is almost always to be seen in the immediate vicinity of Reikevig), I could not help reflecting on the singular situation of our little party of Englishmen, not exceeding in all five or six and twenty persons, removed from all possibility of succour, enemies to the sovereign of the country, and having moreover, made the governor prisoner and exercised dominion over the whole island, yet, nevertheless, living unmolested by a single native, and undisturbed, except by a few, who seemed to have interested motives in falsely representing the people as ripe for insurrection. Our state of security was undoubtedly owing to the willingness of the natives to shake off the yoke of the Danes, and to the full persuasion they entertained that it was the British alone who could supply them, in times of scarcity, with necessary subsistence, and keep them from a state of actual starvation. Of the existence of such a feeling every day's residence at Reikevig furnished abundant testimony; but still more satisfactory were the proofs I received, as well during the present as in my succeeding excursions, when the satisfaction of the inhabitants, at the prospect of being placed under English government, was repeatedly expressed to me, and that, not only by the poorer class of people, but also by those high in power in the island. On my arrival at Reikevig, between six and seven o'clock in the morning, so far from finding the Margaret and Anne in readiness to sail, it was even doubtful whether she would be so during the course of the next week, which was to me, and perhaps to me only, a fortunate circumstance, as it afforded an opportunity of seeing more of the country than I had lately expected it would be in my power to do. I determined therefore, following as well the recommendation of Stiftsamptman Stephensen as my own inclination, to avail myself of the kind invitation given me by his son, the Etatsroed and chief justice of the island, and visit the district of Borgafiord where he resides; but, as the care of my herbarium and the arranging of the other collections made in my late excursion, required two or three days, I was unable to set out before the following Friday, when the Stiftsamptman again insisted upon supplying me with horses, tents, &c. Independently, indeed, of the preservation of my treasures, I had also other motives for thus delaying my journey to Borgafiord: one of them was my wish to be present at the great salmon-fishery, at a river not far from this town, which was to take place on the twenty-fifth of this month; and another was my desire to visit the sulphur-springs of Kreisevig, which Count Tramp had obligingly recommended to my attention, as being amongst the greatest curiosities that the island affords.

Sunday, July 23. We had now been so long in anxious expectation of the arrival of the Flora, another merchant-vessel belonging to Mr. Phelps, which was to sail almost immediately after us, that we began to fear lest some accident should have happened to her on the passage; and I felt myself particularly uneasy on the subject, as I had considerable reason to expect by her my friend, Mr. Borrer, in whose company I had found such pleasure the preceding year, when he participated with me in the fatigues and enjoyments of a tour through the north of Scotland and the Orkney Islands. It may, therefore, easily be guessed how much, in my present situation, when any society would be valuable, I longed for that of a man, whose taste for natural history was congenial to my own, whose friendship I was well assured of, and whose natural acuteness and various information could not fail materially to promote the object we both had in view. We were consequently not a little gratified on having word brought to us at two o'clock this morning, that a vessel was beating into the bay, and that she was, in all probability, the Flora. On a nearer approach we were certain of her being so; but it was not till four in the afternoon, when she came to an anchor, that I had the disappointment to learn from the captain that there was neither Mr. Borrer on board, nor a single letter from any of my friends in England. The vexation of such a disappointment could not but be severely felt; but the additional regret caused by the idea of my being forgotten by those, whose memory I cherished most fondly when separated from them by a such a distance, was done away when I found that the Flora had left Gravesend only two days after us, and had been detained ever since on her passage, which occupied no less than seven weeks.

Tuesday, July 25. This, which was the day* appointed for the catching of the salmon in the Lax Elbe, at a place near its

* It is to be observed, that for a few days previous to this, nets had been placed at the mouth of the river, to prevent the fish escaping to the sea on their return from spawning; besides which, early in the morning of the same day, for some considerable way up the river, other nets were extended across from bank to bank, at intervals of a few yards, with the view of enabling those who are engaged in catching the fish to do it with the greater facility.

confluence with the sea, is held as a sort of annual festival by the natives for many miles round, and afforded a scene of gaiety and pleasure that I should scarcely have expected to witness in Iceland. At ten o'clock in the morning I repaired to the spot amidst hundreds of natives, some on foot, but more on horseback, all drest in their best apparel, and presenting a truly interesting spectacle, to which the unusual fineness of the day contributed not a little. On every side were to be seen the happy countenances of the natives, and there was visible among the different ranks of people a degree of familiarity that is, perhaps, scarcely to be met with in any other country; for men, women, and children, of all ages and conditions, the Bishop, the Etatsroed, the Landfogued, Amptman and Sysselman, the Midwife, the Washer-woman, and the Tailor, were all conversing with each other without restraint, and on terms of perfect equality. The individuals just enumerated, male as well as female, were clad after the Danish fashion; but among the rest, especially the females, the distinction of dress was more striking; for whilst some, in their less ornamented costume, were riding astride upon their horses, those of higher rank, with finer clothes, were sitting in easy and richly-carved side-saddles, holding in their right hand the rein, and in their left, a whip of black leather, prettily variegated with the white quills of the feathers of the eider-duck, which they contrive to intermix in the braids. Seated upon a heap of stones, in one place, was to be seen a cheerful groupe of Icelanders with a bowl of skiur or of butter before them, which they were eating as a relish to the dry but uncooked heads of the cod-fish ; and, at a little distance from them, a party of Danes had laid aside their favorite pipe, and were regaling themselves with slices of smoked salmon, placed between rye-bread and butter, which they every now and then washed down with the contents of their rum-bottle. On arriving at the banks of the river, about six miles from Reikevig, I remarked a numerous party of men and women wading in the water up to their knees or even waists, and catching with their hands the fish which swarmed in the deeper parts of the stream. As soon as caught, they threw them on shore, where another party was employed in counting them and flinging them into wooden panniers, in which they were to be conveyed upon the horses to Reikevig, there to be salted. Mr. Savigniac, who displayed considerable dexterity in seizing the salmon in the river, afforded infinite amusement to his female assistants, who took great pleasure in throwing the largest of the fish at him, and, as often as they could strike him on the head or face, or on any part where the blow would be least acceptable, united in a loud peal of laughter. Far from being ashamed of this little trick, they would wade up to him, assure him of his skill as a fisherman, and, with great familiarity, ask him to shake hands with them. Before three o'clock in the afternoon two thousand two hundred salmon * were caught in the Lax Elbe, all of which Mr. Phelps

* To catch such a quantity as this would be considered as extraordinary, or even wonderful, in any other country. Pennant, speaking of the Scotch fisheries, says, "The miraculous draught at Thurso is still talked of, not less than two thousand five hundred being taken at one tide within the memory of man." *Tour in Scotland,* vol. i. p. 202.

bought of the proprietor of the place, and cured two-thirds of them for exportation *; the remaining third being allotted to those who gave their assistance at the fishery, as a compensation for their trouble.

Wednesday, July 26. At six o'clock this morning Mr. Phelps and I set off for the purpose of visiting the sulphur-springs of Kreisevig, which are about a day's journey distant from Reikevig. The first nine miles brought us to the house of Mr. Sivertsen, at Havnfiord, at which place, the great bed of lava, called Gardehraun, forms a range of cliffs to the sea, close by whose margin masses of lava of vast size are dispersed in such a manner, that a stranger would conceive the passing of them to be scarcely practicable. In other places we were obliged

* In this, as in many other points of view, it is unfortunate for the Icelanders that Mr. Phelps' stay was so short among them; for in former years they have had no means of disposing of the salmon they caught; and, as the exporting of them on their own account has been wholly out of their power, all beyond what might be requisite for their own consumption has been necessarily wasted.

to follow a very devious course, to avoid great holes, of the shape of inverted cones, which had every appearance of being the craters of volcanoes, that had been long since extinguished. Havnfiord contains only two or three merchants' houses and their factories, together with a few peasants' huts scattered about on the small patches of grass that are here and there met with among the hraun, from which, indeed, they are not easily distinguishable; the smaller pieces of that substance composing the walls of the cottages, whose turf roofs only differ from the grassy patches in their superior verdure. A considerable quantity of fish is cured at this place, both for home consumption and exportation. Among the species used for the former purpose is the *Cyclopterus Lumpus*, to the different sexes of which the natives have given different names; calling the male, *Randmage*, a term applicable to it alone, from the circumstance of its having a red belly, and the female, *Grasleppa*, from being grey beneath. This is one of the most hideous of all fishes in appearance, but is highly curious from the nearly circular fleshy appendage on the underside, with which, while alive, it adheres so firmly to whatever it fixes upon, that a pail of water may be lifted up by means of it. The bony ridge on the back of this animal, in all the specimens that came under my observation, was much more elevated than the figures and descriptions both in Pennant and Shaw had led me to suppose I should find it, and added considerably to the general deformity of the creature. To render it an article of food among the Icelanders, nothing more is requisite than to cut away the muscular part of suction, together with a considerable portion of the skin of the belly, and then remove the entrails, which form the greatest part of the bulk of the fish; after which, the small portion of flesh that remains upon the bones is hung up to dry upon the walls of the houses. Bad weather, as in other journies, also accompanied me on this, and the rain, after we had eaten our breakfast, poured down with such violence, and continued so long, that we thought it most prudent to accept Mr. Sivertsen's invitation, and remain at Havnfiord the whole night. Indeed, we were far from considering our time misemployed here, since our host was a gentleman who had twice visited England, and who, from his knowledge of the language of our country, and his excellent abilities, was both able and willing to give us information on various subjects relative to his own island.

Thursday, July 27. At an early hour this morning the rain had not in the least abated, neither was there any prospect of its soon doing so, and we therefore determined to disregard it and proceed to Kreisevig immediately; in pursuance of which, having procured a guide, and being furnished with provisions, at six o'clock we set off, in company with Mr. Sivertsen's son, a young man who could converse with us a little in English, and who kindly offered to attend us. We rode round the head of the Bay of Havnfiord, and continued our route over a very uninteresting and desert country for about six or eight miles, when we came to a part of the great bed of lava which bears the name of Hvassa-hraun *, where, on account of the

* Among this Sir George Mackenzie remarked some lava, which appeared as if it had ascended in its course; which, he says, may be accounted for by the formation

unevenness of the surface, we were compelled to travel a slow foot-pace, and, indeed, to continue doing so almost the whole of the rest of the way. We approached tolerably near the western extremity of the Helgafel * range of mountains, which, though of no

of a crust on the cooling of the surface, when, a case or tube being thus produced, the lava rises in the same manner as water in a pipe.

* Helgafel is remarkable for having had in its neighborhood not only the seat of the court of justice, but also, in early times, a temple of idols at the foot of the mountains. "C'est entre Helgafel et Torsnaes qu'un des premiers habitans du pays vint établir sa demeure. Il était Norvègien, et s'appelait Thorolf-Monstraïskaeg. On avait construit un baillage et un temple d'idoles au pied de la montagne, vers l'ouest, près d'un golfe; ce qui fait que l'on appèle Hofstade, la place et le bâtiment qui existent encore aujourd'hui. On y voit des vestiges des champs et pâturages qu'il y avait alors. Thorolf et ses descendans croyaient qu'après leur mort ils viendraient habiter Helgafel; c'est aussi d'après cette idée qu'ils laissaient jouir leurs bestiaux d'une pleine liberté. Il etait défendu de les faire aller de force, il fallait attendre au contraire qu'il leur plût d'avancer à leur gré: surtout il n'était pas permis de les frapper. Ils regardaient la montagne dont nous venons de parler comme un lieu saint; personne ne s'enhardissait à le regarder qu'il ne se fut lavé la face et les mains. Il en était de même

great elevation, had considerable masses of snow lying on various parts of their bleak and barren sides. Leaving these on our left, we passed between several small insulated mountains, sometimes entering vallies abundantly clothed with *Trichostomum canescens*, and so surrounded on all sides by hills of black and porous lava, that for a short space of time it seemed as if our farther progress would be absolutely prevented. In these situations the elegant *Geum rivale* flourished as in more temperate climates, and *Orchis mascula*, which was equally abundant, produced both reddish and white blossoms. The only birds that we met with were numerous coveys of Ptarmigans, which ran about within a few yards of us without shewing any symptoms of fear. The nearer we ap-

du bâtiment où se tenait le bailliage; ce lieu était comme sacré. C'est ce qui lui a fait donner le nom de Dritskiaer, qu'il a conservé jusqu' à ce jour. Cette soumission trop rigoreuse en elle-même, ne pouvait pas exister long-temps. Les esprits se révolterent, et il survint une petite guerre civile, qui fit que l'on transportât le bailliage plus avant dans les montagnes, à peu de distance d'Helgafel. Cet endroit est situé nord-est, et se nomme encore Thingevalle." *Voyage en Islande, tom.* ii. *p.* 293.

proached to Kreisevig the more broken and uneven the country became, and we were soon within view of some fine black and excessively rugged mountains, which lay between us and the object of our journey, and which we had consequently to cross. At the foot of these we rested our horses for a few minutes, to prepare them for the ascent, which, though steep, was for some way not difficult. At length we approached the brink of a vast hollow, in shape like an inverted cone, the regularly sloping slides of which were composed of loose pieces of rock, while the bottom alone produced a little grass and moss. Into this cavity, which has an Icelandic name signifying *kettle*, it was necessary for us to descend a few yards, after which, turning to the left, we had to go along a track so narrow, that there was no more room than was absolutely required to enable our horses to set one foot before another, on account of the steepness of the ascent on one side, and the suddenness of the descent on the other; till, on reaching the opposite extremity of the place, we ascended to the top, and once more continued our painful journey up the sides of this rocky mountain. In many places, for a considerable extent, the hill had nearly a level surface, upon which were scattered at various distances insulated pieces of rock of immense size, and of the rudest figures, some of them having sharp and apparently vitrified summits, whilst others were rounded off on every side, and had probably rolled into their present situation from the higher peaks of the mountain. Although the singularity of this scenery afforded us no small gratification, our own wet condition, (for the rain still continued unabated,) the excessive cold of these more elevated regions, and the pelting of the great hailstones, which a strong easterly wind drove against our faces, made us rather wish for the shelter of the vallies. On reaching the highest summit, however, we were inclined to forget our uncomfortable situation, whilst looking down into the valley which surrounds Kreisevig. Our view, indeed, was confined from the unsettled state of the atmosphere, yet, at intervals, as the gusts of wind dispersed the clouds, we beheld, in the midst of a green and extensive morass below, three or four lakes, with steep and rocky banks, and, in different parts of the sides of the mountain on which we stood, vast bodies of smoke rising to a great height from the then concealed sources. Our fatigue in descending to the marsh was scarcely less than we had experienced in climbing the opposite ascent; but when we reached the foot and looked to the more elevated parts of the hill, another picture presented itself. The mountains in the range which we had just crossed, for a considerable length of way were black and rugged beyond conception, and jagged upon the upper parts into the strangest figures that can be imagined. Columns of steam were ascending from various places on their sides, especially in the gulleys; some rising near the base of the hills, others almost adjoining the very summit; and the apertures, that gave birth to these columns, also poured out a bolus of different colors, but more especially white, which was conveyed away by the streams of water, and either streaked the hills with party-colored lines, as it descended with them in their devious courses to the plain below, or formed large patches by a deposition of its substance in the hollows of the rock. As our guide was not sufficiently acquainted with the country, to be able to point out the particular objects that were most deserving of our attention, we thought it better to procure ourselves a cicerone to these places from among the inhabitants of a solitary hut, at about two or three miles distance; but still we could not resist the present temptation of alighting from our horses, to visit one of the sulphur-springs that lay in our route. It was situated in a valley, at the foot of the precipice; on entering which, we crossed, with cautious steps, some heaps of bolus, intermixed with incrustations of sulphur, and arrived at the edge of the fountain, where, in addition to a whitish and turbid water that was thrown out to the height of two or three feet from an aperture of no small dimensions, we found a muddy paste oozing from other orifices at various distances. All of these sent forth great clouds of steam, which, together with the sulphureous exhalation that was wafted about by the wind in different directions, frequently obliged us to shift our situations. It was in endeavoring to avoid one of these unpleasant gusts, which threatened to annoy me while I was gathering some specimens of the mineral productions of the place, that I jumped up to my knees in a semi-liquid mass of hot sulphur and bolus, in which I should probably have sunk to a considerable depth, had I not instantly thrown myself with my whole length upon the ground, so as to get my hands on a more solid soil; by means of which I dragged myself upon terra firma, and relieved the anxiety which those who saw the accident were entertaining for my safety. An unusual quantity of cloathing about my legs prevented my experiencing any other ill effects from the heated mass than a sensation which was rather uncomfortable than painful, and was not of long duration; so that, after being well scraped from a substance that attached itself like cart-grease, we proceeded on our way. In the midst of an extensive swamp we passed a lake*, with steep and rocky banks, whose waters surprised us not less by their excessive

* It is of this lake that mention is made in the *Voyage en Islande,* where it is observed, *tom.* v. *p.* 58, " Le lac Groenavatn, près la soufrière de Kreisevig, est remarquable d'abord par la couleur verte de son eau, qui provient probablement de sa profondeur, et ensuite, par les

clearness than by their deep bluish-green tint. The sky was clouded, nor was there any thing to be discovered on the shores that could reflect that color, for which we could therefore account by no other means than by supposing that a bottom of greenish bolus had imparted its tint to the waters. The numerous shallow pools scattered about the morass neither possessed the hue nor the clearness of the lake, but were strongly impregnated with the sulphate of iron. An hour's ride in this marsh, nearly mid-leg deep in water and among abundance of *Betula nana*, brought us to the residence of the inhabitants of Kreisevig, by whom we were ushered into a low turf building, which, though small, and much incommoded with dirty clothes, stockings, saddles, &c., afforded room for our little party to take some refreshment, and proved a most welcome shelter from the unceasing inclemency

relations que les riverains font, des créatures singulières qui doivent s'y trouver, et qui se montrent quelquefois un instant au-dessus de l'eau. Une personne nous assura avoir vu un petit monstre de conformation approchant d'un marsouin, mais qui disparut presqu' aussitot qu'il parut."

of the weather. As soon as we had recovered the use of our eyes, which the almost total darkness of the place for a time deprived us of, Mr. Phelps expressed his astonishment at seeing, upon a sort of table, two large candles, articles of extremely rare occurrence in this country, and these, also, placed in brazen candlesticks: he began, therefore, to suspect that we must be in the house of some man of property, in spite of the quantity of dirty apparel that, hanging from the beams, seemed to persuade to the contrary: nor could he for some time be induced to credit my assurance that the place where we were was no other than the church of Kreisevig; that the table we leaned on was the altar, and the two candlesticks its constant appendages. It is a frequent custom with Icelanders, whose dwellings are in the vicinity of a church, to receive their guests in it, as affording a more spacious and convenient apartment than any of their own; and such was the case even here; though in this edifice, except the light admitted by a small door, a little aperture in the wall above the altar, about six or eight inches square, was all that answered the purpose of a window. Here, however, we were regaled with some excellent sheep's milk, and, having urged our request to our host that he would accompany us to the sulphur-springs, we, after a short rest, again mounted our horses. Although in the vicinity of a remedy so noted for the cure of a certain disagreeable cutaneous complaint, we observed, by the swellings on the hands of our Kreisevig guide, and by his incessant scratching, that he had not, any more than some other people whom we saw living near the sulphur-springs, made such use of it as would be done in other countries; but, on the contrary, it rather appeared that the disorder was here more than usually prevalent. The first place to which he led us was a spot about two miles from the village, where a thick and muddy water was boiling up from a number of small orifices, occupying a hillock, of some yards in extent, but composed entirely of *Bolus* * of various colors; among

* It may be well to observe that *Bolus* is described by mineralogical writers as a viscid earth, less coherent and more friable than clay, more readily uniting with water, and more freely subsiding from it. It is soft and unctuous to the touch, adheres to the tongue, and by degrees melts in the mouth, impressing a slight sense of astringency.

which, however, red was the predominant one: a bluish grey, also, was extremely abundant, and we met with yellow and yellowish white in smaller patches; all of them extremely soft and unctuous to the touch. The boles of different colors, although not divided from one another by the intervention of any other mineral substance, were in general unmixed, and, by digging to the depth of a foot, we were enabled to see them lying in separate strata, each color being kept quite distinct from the other. In Iceland the only bolus that the natives make any use of is the red, which mixes with oil, and is frequently employed by people of higher condition to color the wooden doors and the entrances of their houses. I have also seen tables painted with this ingredient, which, in this country, where paint of any kind is scarcely ever seen, seemed to me to look extremely well. From these beds of bolus we proceeded towards a fountain of considerable dimensions some way up the side of a mountain, passing, as we went along, numerous others of less importance, most of them environed by bolus and sulphur. Of the latter substance, the spring, that we were now approaching, produced the greatest quantity, and the finest specimens, I believe, in the island. We rode some way, till the softness of the earth beneath caused the horses to sink too deep to render it prudent to continue that mode any longer, and we therefore left our steeds, proceeding onwards, as far as it was by any means safe to venture, with the utmost caution. The appearance of the surface is often very deceitful; for, when it seems most firm, a thin indurated crust of crystallized sulphur * and bolus not uncommonly con-

* Volcanic soils in many parts of the world produce sulphur in greater or less quantities. I have not, however, read of its being found any where in such abundance as in the province of Satzuma, in Japan. "It is dug up," says Kæmpfer, in his history of that singular country, "in a small island, which, from the great plenty it affords of this substance, is called Iwogasima, or the Sulphur Island. It is not above an hundred years since the natives first ventured thither. It was thought before that time to be wholly inaccessible, and by reason of the thick smoke, which was observed continually to arise from it, and of the several spectres, and other frightful uncommon apparitions, people fancied to see there chiefly in the night, it was believed to be a dwelling-place of devils, till at last a resolute and courageous man offered himself, and obtained leave accord-

ceals a considerable mass of the same materials in a hot and almost liquid state, so that we literally walk "per ignes, suppositos cineri doloso." This kind of soil became still

ingly to go and examine the state and situation of it. He chose fifty resolute fellows for this expedition, who upon going on shore found neither hell nor devils, but a large flat spot of ground at the top, which was so thoroughly covered with sulphur, that wherever they walked, a thick smoke issued from under their feet. Ever since that time this island brings in to the prince of Satzuma about twenty chests of silver per annum, arising only from the sulphur dug up there.—The country of Simabara, particularly about the hot baths above mentioned, affords also a fine, pure, native sulphur, which, however, the inhabitants dare not venture to dig up, for fear of offending the tutelar genius of the place, they having found upon trial that he was not willing to spare it."—The Kamtchadales, as well as the Japanese, have a dread of the hot-springs in their country, arising from a similar supposition that they are the abode of demons. Thus, speaking of the boiling fountains of Opalski, or Osernoi, situated nearly midway between the Lopatka and Bolshoiretsk, Martin Sauer observes, that the Kamtchadales suppose them to be the habitations of some demon, and make a trifling offering to appease his wrath; without which, they say, he sends very dangerous storms. See the *Account of an Expedition to the Northern Parts of Russia, by Commodore Billings,* p. 303.—There is also in Arabia a tradition about a hot-spring, near Suez, that the jews

more and more dangerous the nearer we approached to the spring, and, indeed, prevented our being so close to it as we wished. An elevated rim, about two feet high and three feet in diameter, composed of a dark bluish-black bolus, formed a complete circle round the mouth of the spring, the water in which was sometimes quiet and sunk about two feet in the aperture: at other times it ejected with great noise a turbid and blackish liquid to the height of from five to seven feet. At all times clouds of steam, strongly impregnated with sulphureous exhalations, were issuing from the aperture, but during an eruption of the waters the quantity of both was very considerably augmented. The view of this spring, from a little lower down the mountain, together with the surrounding scenery, had an effect the most extraordinary that can be conceived. From the dark

passed that way, and Pharaoh's army was drowned there, which has caused the place to receive the name of Birket-el-Faraun. The Arabs imagine that Pharaoh is doing penance at the bottom of this well, and vomits up the sulphureous vapor with which the water is impregnated." *Niebuhr's Travels*, in *Pinkerton's Collection*, vol. x. p. 8.

colored and elevated margin of the fountain extended for a great way in every direction the yellow crust of crystallized sulphur, raised into a gently swelling hillock by the soft bolus of unmeasurable depth beneath; and from the centre of this trembling mass a crater was vomiting forth, with a tremendously roaring noise, to the height of four or five feet, a thick blackish liquid, accompanied by vast bodies of steam, which now ascended perpendicularly, and now were driven down the sides of the hill by the frequent eddying gusts of wind which issued from the chasms that abounded in the neighborhood. A back ground, worthy of such a picture, was supplied by the dark and rugged sides of the mountain that, extending all around, formed a chain of rocks, which, in addition to the rudeness of their figure, were the most barren that can be imagined. A few lichens and mosses alone broke the uniform blackness of their surface; and these, far from being in a luxuriant state of vegetation, were scarcely to be discerned at a little distance, and appeared only minute greyish spots. How unlike to the volcanic scenery of this frigid region must be that of

Ætna, where, according to the account of an ingenious traveller *, "every beauty and every horror are united, and all the most opposite and dissimilar objects in nature; where in one place you observe a gulf that formerly threw out torrents of fire, now covered with the most luxuriant vegetation, and from an object of terror become an object of delight; where you gather the most delicious fruits and tread upon ground covered with every flower; where you wander over these beauties and contemplate this wilderness of sweets without considering that hell, with all its terrors, is immediately under your feet; and that but a few yards separate us from fire and brimstone." The horrors alone of the picture given us by Brydone are to be met with in the volcanic mountains of Kreisevig: for luxuriant vegetation, fruits, and flowers, other countries must be searched, and yet, in spite of the absence of every beauty that could attract, or excite a pleasurable sensation, I doubt whether a traveller ever turned his back upon Ætna with more regret than we felt

* See *Brydone's Tour through Sicily and Malta*, p. 93.

when we quitted the strange but desert scenery of this place. To myself, indeed, the regret was no more than the being deprived of the power of beholding one of the most awfully impressive scenes that the world can furnish, or even imagination can conceive; but not so with my companion, who had hoped that it might have been possible to have met in the sulphur-springs with an article of commerce that might at once have been highly advantageous to himself, and beneficial to his country, but who now found to his extreme vexation that, small as is the distance of Kreisevig from the sea, the obstacles interposed by the nature of the intervening country were such as forbade the idea of a commercial speculation. To have collected it in a place where the population is so thinly scattered, would have been attended with very great expence; and to have conveyed it on horseback over so rocky a tract as lies between Kreisevig and the nearest harbor, would have been almost impossible; and I therefore read with surprise, in Horrebow, that early in the last century the gathering and exporting of it were objects of considerable advantage to the natives. Myvatn, in the more northern part of the island, is said to be almost the only place, except Kreisevig, where this mineral is produced in any considerable quantity. We were the more vexed at being obliged to return, because the incessant rain prevented our bringing away any sketch of a spot, of which words can give but a very inadequate idea, and which is in itself alone a sufficient recompence to a mind even the most incurious, for the fatigues and privations necessarily attendant upon the travelling about Iceland. On our way back to Havnfiord, by the same route as we went in the morning, the most interesting occurrence to me was the meeting with *Parmelia sarmentosa* on the rocks of lava in some abundance. A little after midnight, wet and weary, we reached Mr. Sivertsen's house, and on the following morning returned to Reikevig, with our horses no less exhausted than ourselves, and mine so lamed by the beds of Hraun, that I was never after able to make use of him.

Friday, July 28.

Saturday, July 29.

Having been informed that when travelling, as I purposed to do, loaded with much luggage and tents, it would require three days before we could reach the house of the Etatsroed, at Inderholme, in the district of Borgafiord, and that a portion of this time must be allowed for the horses to rest, I thought it best to make the present day's journey extend no farther than the foot of the mountain Skoul-a-fiel, which lay in our route, and afforded pasture for the horses, besides offering to myself the opportunity I wished of devoting the whole of the next day to the examining of the hill and its neighboring chasm. The fineness of the morning afforded me great pleasure, and, as the wind had veered to the north, I looked forward to a few days of bright and dry weather. Horses and guides having been furnished me on the preceding day by the Stiftsamptman, I sent them forward in the early part of the morning with the baggage and a week's provisions of ship's stores, giving them directions where they should pitch the tents, in case they arrived at the journey's end before we should come up with them. Mr. Phelps, by kindly permitting Jacob to accompany me a second time, did me an essential service, as the fidelity and honesty as well as the good sense of this man rendered him an useful servant, and often an amusing companion. The various climates he had visited, and the hardships he had suffered, from his earliest youth, enabled him to endure alike heat and cold, and to bear the greatest fatigue without ever uttering a single complaint. In his broken English he not unfrequently relieved the wearisomeness which attended travelling over the long and dreary moors of Iceland, by relating the adventures that he had met with in his many voyages and travels, particularly in a journey that he had made from Petersburgh to China. By birth he was a German, but could talk English and Danish, and, besides acting as interpreter, he was of considerable use to me as a butcher, as well as in cooking, and occasionally in washing for me *. I certainly experienced great inconvenience from my ignorance of

* These few remarks, which I have thought due to the short but faithful services of this man, were scarcely written down (July, 1810), when I received from Mr. Phelps the unwelcome intelligence, that he was no more. A vessel from Iceland brought the information, that he, together with another of the crew, who after the loss of

the Icelandic language, as, except in a very few instances, I could only obtain information from the natives through the medium of two interpreters; my question being put in English to Jacob, who translated it into Danish to my Reikevig guide, and he, again, in Icelandic, made it intelligible to the person I wished to address. The answer, also, was necessarily returned by the same circuitous way. It was half past six in the evening, before Jacob and myself set out, when we travelled as fast as the roads, which are better in the immediate vicinity of the capital than almost any where else, would permit us; stopping only to admire, and to gather specimens of, the elegant *Saxifraga Hirculus*, which adorned, in the greatest profusion, the numerous springs of water that we met with near our road. It was in this journey, for the first time in my life,

the Margaret and Anne, had remained at Reikevig, and married and settled there, had gone out one day to sea on a shooting excursion with Mr. Savigniac, when the boat was unfortunately overturned, and the two sailors perished. The body of poor Jacob was thrown on shore the next morning, but that of his companion had not been found.

that I saw its beautiful yellow blossoms, and I thought I could never gather enough of the plant. In about three hours we overtook our luggage horses and guide: despising, however, a conductor in a tract of country, over which we had twice travelled before, we hastened forward on our way, but had scarcely lost sight of our company than we saw reason to regret our precipitancy; for we found ourselves so encompassed by bogs, that we were at a loss how to proceed. In urging my own horse through a swamp, he floundered and threw me, and I had great difficulty in extricating him from his unpleasant situation. Jacob, by a more circuitous route, reached me in safety, and we continued our journey till about ten o'clock, when we arrived at the foot of Skoul-a-fiel, and fixed upon a little verdant plain by the banks of a wide and extremely rapid torrent for the situation of our tents, which did not come up to us before twelve o'clock. At about half a mile from us was a peasant's house, called, if I recollect right, Skykeaster, to which I dispatched Jacob, according to my usual custom, for some fuel to boil our kettle and some milk *. In the owner of this house, for the first and only time in

* For the convenience of having the milk brought down to me, I always sent bottles to the cottagers; but it never came into my mind to inquire what means were employed to convey the fluid into such a vessel from the large and shallow dishes in which it is kept by the natives; in a country, too, where funnels cannot be supposed to be in use among the poorer class of people. I should, probably, to this day, have remained in ignorance of the method, had I not, a little previous to my leaving the country, been informed, as well by the Danes at Reikevig, as by some natives (persons worthy of credit, and whose names if necessary I could now mention), that the milk is first taken into the mouths of the women, and then spirted into the bottle.—Let it be remembered, that I do not mention this circumstance as one to which either Jacob or myself was a witness, neither could this well have been the case, for the bottles were always carried into the house by the women, and returned to us filled; but, from the respectability of my informers, and the simplicity of the mode, it really appears deserving of credit.—Linnæus, on the Lapland Alps, partook of *Misseen*, a kind of whey, under circumstances equally filthy. "Its flavor was good," he observes, "but the washing of the spoon (which was done by spirting water upon it from the mouth) took away my appetite, as the master of the house wiped it dry with his fingers, whilst his wife cleaned the bowl, in which milk had been, in a similar manner, licking her finger after every stroke." *Lach. Lapp.* vol. 1. p. 293.

the island, I met with a deviation from that genuine hospitality which so strongly characterises the inhabitants of Iceland. In all my other excursions I was furnished with milk, fuel, or whatever the house afforded, with the greatest cheerfulness, and with the strongest marks of welcome; and, even if I remained for some days in one spot, I never thought of making a return, except it was in the trifling articles of snuff and tobacco, until I was about to take my departure from the neighborhood. It is therefore as a single instance of avarice and mistrust that I mention the owner of Skykeaster, who, on coming down to my tent with a few birchen twigs that were not sufficient to boil the kettle, and about a pint of milk, demanded two marks and eight skillings *. This I paid him immediately, letting him know at the same time that, had his conduct been different, he would have been better recompensed; at which he was so much vexed that he offered to return the money, and furnish me unconditionally with as much more of the milk and fuel as I wanted. A strong

* About one shilling and eight-pence of our money.

northerly wind, which rushed down the gullies of the mountain, made us regret the not having fixed upon a more sheltered spot for our habitation during the night, and I therefore forded the river, in hopes of finding such an one nearer to the foot of Skoul-a-fiel; but our own fatigue, the weariness of our horses, and the difficulty that would have attended their conveying the luggage over the excessively rocky bed of the river, induced us to prefer accommodating ourselves as well as we could to our present station, trusting that, by fixing the tent-pins deeper in the ground, and placing our luggage-saddles, &c. round the bottom of the tent, we should be able, at least in some degree, to keep out the wind and cold. Scarcely, however, had we composed ourselves upon our homely bed, when a most violent blast tore up the pegs, and exposed us to the utmost fury of the elements. In vain did we again attempt to fasten them: as often as we flattered ourselves they were secure, the force of the wind immediately drove them out, and the intense cold, added to the continual flapping of the canvass with a noise like thunder, prevented our en-

Sunday, July 30. joying a moment's rest. Very early in the morning, therefore, of the following day, I hastened to the river, designing to pursue its course, with a view of entering at the deep chasm, and proceeding along it to the perpendicular column of rock, which I had previously remarked on my return from the Geysers. The stream, for some way, ran through a tolerably level country, but, in proportion as I advanced, its banks became more precipitous and rocky, and continued to increase in elevation and grandeur, so that, not unfrequently, nothing more was to be seen than the steep and craggy cliffs which arose to a great height on each side of me, and the impetuous torrent that ran foaming between them, scarcely leaving a narrow ledge that might afford room for my feet, and repeatedly tumbling in its passage over shelves of rock, thus forming cataracts, which varied in height from two to three and even ten feet. Occasionally, however, a cleft in the northern side brought to my view the lofty top of Skoul-a-fiel, with its pointed summits, looking as if it took its rise from the very edge of the precipice. At length my farther progress was stopped by the rocks closing in so much as to leave room for nothing at their base but the narrow and furious course of the river. It was near this spot that I found both *Fontinalis squamosa* and *falcata* full of capsules, in a deep pool among the rocks, and mixed with them was also *Rivularia angulosa* in some plenty. The rocks in a steep ascent, which I climbed in order to reach the top of the chasm, produced an *Epilobium* which was not yet in blossom, but appeared, from its broad and glaucous leaves, to be undoubtedly new to me. *Veronica fruticulosa* was here in full flower, and some unknown *Salices*, likewise, rewarded my morning's excursion. Fearing lest I should not have sufficient time to ascend Skoul-a-fiel, if I proceeded any farther in the same direction, I returned to the tent, and, after a hasty breakfast, set off on horseback with Jacob, on our way to the mountain. We forded the river, and afterwards climbed a steep but grassy hill, whose swampy summit afforded me some fine specimens of the rare *Splachnum vasculosum*. On descending by the opposite side, and crossing another stream, we came to the base of the mountain, up the precipitous sides of which we mounted in a diagonal direction, keeping in a beaten track for some way, but at length directing our course, in the nearest line, for the highest summit. We were soon compelled to leave our horses; for, though the base of the hill had been firm rock, interspersed with a few patches of vegetation, we shortly came to a part, from which, to the very peak, the whole was altogether composed of small loose pieces to the greatest degree barren and desolate; except in those little spots in which the *Trichostomum* had formed a bed, and retained a sufficiency of moisture to supply with scanty nourishment a few miserable specimens of *Salix herbacea* or *Silene acaulis*. It is hardly possible for any person, unless from experience, to form an idea of the fatigue of climbing a mountain like this: wherever we placed our feet a vast number of pieces of rock were immediately set in motion, and rolled for a considerable way down, causing us to lose nearly as much ground as we gained, and as often as, to use the words of Dante, by reason of the greater steepness of the ascent,

> "Tra le schegge, e tra' rocchj dello scoglio
> "Lo piè sanza la man non si spedia,"

so that we were compelled to make use of our hands in addition to our feet, these latter were sure of being annoyed by a torrent of the sharp and angular stones striking against them. At length, drenched with perspiration from the violence of the exercise, we reached a ridge of the mountain, which led by a gradual ascent to the summit; great masses of snow lying every where scattered about its precipitous sides. No sooner had we attained this ridge, than we found ourselves on a sudden exposed to the force of the wind, that, driving over some distant snow-mountains, froze us with cold, and at the same time, from its excessive violence, made it prudent for us to sit down, rather than stand, while we surveyed the extensive tract of country that lay, like a map, spread out beneath our feet. To the north was a wilderness of mountains, many of which far exceeded in height the one upon which I stood, and most of them were thickly clad with snow. In the north-west, the most striking feature was Snöefel Jökul, which, taking its rise near the sea, at the western extremity of the syssel of that name, towers to an elevation of not less than seven thousand feet. Its distance from me was between sixty and seventy miles, and I now, almost for the first time, beheld entirely free from clouds this immense rock, appearing like a huge cone of solid snow. The extensive Bay of Faxa-fiord was bounded on the south by the narrow neck of land, called Guldbringue Syssel, producing many mountains of wild and singular forms, springing from among its numerous beds of hraun. The town of Reikevig was plainly to be seen; as well as its harbor, spotted with the vessels lying at anchor, and the numerous little islands. In the south, the eye wandered over a wide tract of rocky moor, beyond which the distant Helgafel mountains varied the line of the horizon. We had scarcely time sufficiently to admire this scene, when, on looking upwards, we saw approaching us a thick cloud, which, covering the summit of Skoul-a-fiel, rolled down the sides, accompanied by gusts of wind, still heavier than before, and soon enveloped us in so dense a fog that we could not discover each other even at a very few yards distance. We continued, however, to ascend by the assistance of the compass, and, when the mist had, for a short time, cleared away, we had the pleasure of finding that we had varied but little from our proper course. The nearer we approached to the summit, the more steep we found the ascent, and the more narrow the ridge along which we had to pass; so that I was glad to be able to assist myself in climbing, by laying hold of a few pieces of rock, which, projecting here and there from among the loose ones, seemed to be still in their primæval state. They lay in strata or laminæ which were easily detached from one another; each stratum being vertical, and not more than one or two inches thick. The whole was of a reddish-yellow color, variously marked and spotted with white, green, and red, so as to have a very beautiful appearance. The highest summit was so much peaked, that it would scarcely afford standing room, even in calm weather: and therefore, with the hurricane which now blew, I was happy at being able to reach it upon my hands and knees, and then, laying myself down upon the sloping side of the ridge, to look over the northern precipitous edge, and view in safety the rapid motion of the clouds passing towards the place on which I was, across the valley which separated this from other mountains. The superior height of Skoul-a-fiel above all those in its immediate vicinity caused it to attract these clouds more than any of the rest, yet the violence of the wind did not suffer them to remain long upon it, but soon dispersed them after they had rolled a little way down the southern side. Vegetation here was very scanty: in such places as were free from snow, and lay in their original strata, were to be found *Salix herbacea, Saxifraga oppositifolia, Polytrichum sexangulare* and *Lichen geographicus.* Nothing could be more easy than our descent among the loose stones, where the principal requisite was to be well provided with stout shoes, and we therefore soon got under shelter from the wind. On our right was a deep ravine, from the bottom of which arose a spring that supplied a little stream, which I resolved to take in my way, in order to see what plants it produced, while Jacob hastened forward in search of our horses. Here I spent some time in collecting one of the richest botanical harvests I ever made in one day. Some grasses, a *Veronica,* and a *Gnaphalium,* with five or six mosses were all new to me, and I also found several scarce plants that I had never before seen, though I met with them in other places afterwards; so that with these I not only completely filled two vascula and my game bag, but at length began to apply my pockets to the purpose of containing my specimens. On my return I found Jacob fast asleep more than half way down the mountain, holding in his hand one of the horses by the bridle. Having mounted our beasts, we made the most haste we could to our tent; and, as it still appeared possible, before the dusk of the evening came on, to go in search of the rock in the chasm which I failed of finding in the morning, I set out a second time for the purpose, and, keeping above the chasm, was not long before I came within sight of it: this, fortunately for me, happened near a spot where I was enabled to descend to the very banks of the stream, and procure a good view of this remarkable place. The lofty column of rock was entirely separated on one side from the opposite perpendicular wall of the chasm, to which it was on the other side united merely for a few feet from its base, so that the water did not altogether surround it, though sufficiently so to give it a most remarkable appearance. In the faces of the chasm were several basaltic pillars lying in a horizontal direction, firmly imbedded in the solid rock, resembling those figured in *M. Bory de St. Vincent's Voyage* *, excepting only that the Icelandic ones did not extend to the base of the rock, but merely occupied a few yards of the surface. The singularity of this place detained me till a late hour; yet, in spite of the fatigue of the day, I had the vexation to find on my return to the tent, that the continuance of the wind and cold caused me to spend as sleepless and uncomfortable a night as the preceding one.

* See plate xi. of that work, where, on the left hand, is represented a rock containing similar horizontal pillars. Basaltic columns of the kind figured on the right hand of the plate are not uncommon in Iceland.

Monday, July 31. After having given up the early part of the morning to the preservation of my botanical riches, we set off upon our journey, proceeding for the first part of the way over a shoulder of Skoul-a-fiel, and then over a mountain called Swein-a-scaur, the descent of which, through a gulley where we had to cross a torrent at least twenty times, was excessively steep and rocky, and so exposed to the fury of the north wind that we were compelled to alight from our horses and walk. The ground we trod upon was, however, not altogether bare of vegetation; for several grasses and other plants appeared in the places that were free from snow, and at a great elevation *Geum rivale,* which is found in the flat meadows of Norfolk, was no less abundant than its alpine neighbors, *Veronica fruticulosa* and *Arabis alpina.* In some hollows of the rock, that were filled with the water of the torrent, I met for the first time in my life with the *Harlequin duck (Anas histrionica)*, which, from what I could learn, does not seem to be a scarce bird in Iceland. A very serpentine course, in consequence of the steepness of the hill, at length led us into a rather extensive level tract of country, bounded on all sides by black and lofty mountains. For some way near the banks of a wide stream in the centre of this, a tolerable pasture was afforded for our horses, and we rested ourselves awhile at a dwelling called Meurawatl; a thing the more necessary, as a dreary mountain ride lay before us, and we were told it would not be in our power to meet with grass again till we had got round the head of Hval-fiord (the bay of whales). The steep and barren sides of Renewaltehauls afforded nothing interesting, but from the summit the distant view of snow mountains in the more northern part of the island was most grand. Here we rode over a bed of rock, curled on the surface, which, though cracked in a few places with deep fissures, had the appearance of being a solid mass, and of having suffered no change; but not so with a heap of rocks, broken, indeed, yet still of immense size, which, piled one over another to a great height on our left, seemed to have been at a distant period thrown out in a melted state from a volcano, and to be still suffered to remain a monument of some dreadful eruption. Their texture was in parts solid, in other parts porous, their color a brownish-black, speckled throughout with innumerable small white pieces of quartz, which, on a close inspection, had a very pretty appearance. From the dismally barren scene before us, we soon came to a little plain, where the *Bartsia alpina* in full flower made amends for the absence of more abundant and more varied vegetation; but hence to the margin of the water was a dreary scene of abrupt precipices, rugged hills, and rocky streamlets. A river, at the head of Hval-fiord, in discharging its waters over the perpendicular face of a rock, formed a fine cascade, just beneath which, and exposed to the full effect of its tremendous roar, we had to ford the stream, after which, for a few miles, we travelled along by the north side of the lake among heaps of fragments that had fallen from the steep hills, till, about ten o'clock, we had once more the satisfaction of seeing a green spot, which had induced a peasant and his family, after the manner of the ancient Germans *, to fix in it their solitary dwelling. The singular custom which prevails throughout Iceland of giving a name, as of a parish †, to a solitary hut, or at most to the residence of a more wealthy farmer and the cottages

* It is impossible to avoid being struck with the similarity of part of Tacitus' description of the manners of the Germans, to the present rude and simple state of the inhabitants of Iceland, who are compelled from the scantiness of vegetation thus to imitate the people of former days in the distant situation of their dwellings from each other. "Nullas Germanorum populis urbes habitari, satis notum est; ne pati quidem inter se junctas sedes. Colunt discreti ac diversi, ut fons, ut campus, ut nemus placuit. Vicos locant, non in nostrum morem, connexis et cohærentibus ædificiis: suam quisque domum spatio circumdat, sive adversus casus ignis remedium, sive inscitia ædificandi."

† The land in Iceland, at least by far the greater part of it, belongs to the King of Denmark, and a native is at liberty to pitch upon any waste that may suit his convenience, and fix his abode there: his farm or habitation he calls by some name, either taken from the peculiarity of situation, from some neighboring mountain or river, or after himself; "Ut hac ratione," as the learned Arngrim Jonas observes, "primos incolas επωνυμους ipsa loca vel solis nominibus apud omnem posteritatem loquerentur."

of his dependants, will easily account for the crowded names of places which we see in the best maps of the island, and which might lead to a most erroneous idea of its present or former population, unless accompanied by the explanation that in the greater number of instances they are to be understood as the appellations of mere farms*, and never of what in England would be entitled to be called a village. The present place, which, if I recollect well, bears the name of Farit, stands in a singular and interesting situation; being near the head

* "Lands are here divided into estates, which are never subdivided, and are held in three different kinds of tenure:

"King's Land,
"Church Land, and
"Freehold.

"*King's land* is given by the Landfogued to whomsoever he pleases, and the family who occupy it possess it as long as they have an heir and can pay the rent, which is very small, and a tax of one rix-dollar per annum.

"*Church land* is given away by the Bishop and Amptman, and held in the same manner.

"*Freehold* is as in other countries, each estate paying one rix-dollar per annum to the King, in lieu of land-tax." *Sir Joseph Banks' MSS. Journal.*

of Hval-fiord, so that from it we had a noble and extensive view of this arm of the sea, on which were innumerable quantities of the black divers *(Colymbus Troile)* and many flocks of swans. From the agitated surface of the water the violence of the wind raised great bodies of spray, which were driven, like a dense mist, into a valley that opened to the south. In an opposite direction, and near the extremity of the fiord, a mountain of no great elevation afforded us a curious spectacle of another kind; for here a cloud of snow, which was passing nearly over our heads in an unbroken mass, being impeded in its progress by this hill, in a few seconds of time enveloped in a white covering, as with a sheet, its previously brown and barren sides, for nearly half way down. Our encampment was fortunately provided with sufficient shelter from the storm by a lofty and perpendicular rock,

> "Huge as the tower, which builders vain
> "Presumptuous piled on Shinar's plain,"—

the whole so strange in form, and so broken into recesses and projections, that fancy might here, with more justice than in any thing else I ever saw of the kind, picture to itself a heap of vast and ruined towers, placed upon the top of a sloping bank,

> "Whose rocky summits, split and rent,
> "Formed turret, dome, or battlement;
> "Or seemed fantastically set
> "With cupola or minaret.
> "Wild crests as pagod ever decked,
> "Or mosque of eastern architect."

The dreary solitude, and the storms and snow of Farit, did not in the smallest degree prevent the inhabitants from exercising their wonted hospitality. On the contrary, the women here, as at other places, came around us immediately on our arrival, and with a kindness peculiar to the sex inquired into our wants, and offered us all that their circumstances would enable them to afford. As a mark of affectionate good-will, which those most conversant with females can best appreciate, they presented to us their little children to be kissed, and when, as was too often the case, our more refined notions of cleanliness prevented us from profiting by their intended kindness, they begged that we would allow them to kiss our hands, which they did in the most respectful manner, bowing at the same time. The mothers and the elder girls brought to our tent abundance of cream, skiur, and fuel, and pressed us to accept them with such evident marks of earnestness, that it was impossible but that the pleasure which gleamed in their countenances should be reflected in ours, nor could we have failed to have received with satisfaction presents of a far less acceptable nature than those now set before us, if offered by people so situated, and with such hearty good-will; for, to use the words of a favorite poet of the present day,

"*His* gift shall ne'er be scorn'd, who freely gives his all."

Tuesday, August 1. The vain hope of being able to shew to my friends in England some sketches of the rocks of Farit kept me on the spot till nine or ten o'clock this morning, and it was nearly one when we arrived at a farm by the shores of Hval-fiord, where a man announced himself as the servant of the Etatsroed, and added the information that he had been sent out the day before to meet us, and to accompany us to Inderholme. Leaving, therefore, my Reikevig guide with the luggage and other horses, to follow us at leisure, Jacob and myself mounted some steeds sent by the Etatsroed, and hastened forward till we came to the foot of Akra-fiel, a mountain of some height, which rose at no great distance from this gentleman's house, but was separated from it by a morass* that was not to be crossed without much difficulty. In the worst places were laid sod and large pieces of rock, which had been procured from a considerable distance, but, although these prevented the horses from sinking deep in the mire, they by no means rendered the passage firm: yet did this

* Let it not be regarded as a proof of the indolence of the Icelanders, or as setting their characters in an unfavorable light, that these morasses are to be seen, occasionally, in the neighborhood of the best of their houses, and that the roads, not unfrequently, lead over them. All this is, unfortunately, ascribable to the country itself, which is little else than rock and bog; the latter, of so wet and spongy a texture, that no materials, however adapted to the purpose, and no quantity of them, however large, would be sufficient to overcome their stubborn nature, or to make them properly passable.

trackless swamp lead to the very best house in the island, the residence of a man, at once a Danish counsellor of state, and the chief justice of Iceland; one, too, whose talents and acquirements would render him the ornament of any society, but who lived here shut out from all connexion with the literary world. In such of the out-buildings of the Etatsroed's house, as first came in view, was evident a degree of neatness as to workmanship, of elegance as to form, and of regularity as to design, which I had never before seen in the island, and on approaching the door of the principal building, it seemed as if I was actually transported to another country. In point of architecture and materials, it was, indeed, built in the true style of an Icelandic dwelling, and totally unlike the Danish ones of Reikevig, but there was, nevertheless, even in the turf walls and numerous roofs, an appearance of refinement which I little expected to have met with; while the painted doors and the large glass windows were quite novelties. To comfort and cleanliness in the persons of the natives I had not been much accustomed, and was, therefore, the more glad to find them here: for a hearty welcome I was fully prepared; it was no more than I had every where experienced; but those only who have been long exposed to the accents of a language, with the meaning of which they are wholly unacquainted, can conceive how sweet such a welcome sounds, when given me, as here by the Etatsroed, in my native tongue. We entered by a long passage, with a boarded floor and wainscotted walls, and, after crossing another smaller one, arrived at the library, a room of moderate size, well stored with books; adjoining to which was the parlor, which, if I recollect right, had stuccoed walls, painted of a blue color, and a boarded roof and floor. A Danish sofa and other good furniture much resembled such as we have in England, and some ordinary prints, among them one of the Emperor of the French and by the side of it another of the Hero of Trafalgar, served to decorate the walls. Shortly after our arrival, rum with white wine and Norway biscuit were handed round, and, as there was but little time before dinner, we amused ourselves in the library, where I was shewn several valuable and interesting works, relating to the ancient history of the island, as well in manuscript as in print. There were here, also, many of the Latin and Greek classics, and of the most esteemed authors in the German, French, Swedish, and Danish languages, besides, what gratified me more than any thing else, a considerable number of our best English poets. Here, too, I was shewn a translation of *Milton's Paradise Lost* into Icelandic verse, the performance of a priest who had lived in the eastern part of the island, but whose name I cannot now remember. The Etatsroed, who was capable of reading the original, did not express himself at all satisfied with the translation, and I have no doubt of his being a competent judge of the subject, having himself, with much *eclat*, turned into Icelandic poetry *Pope's Essay on Man* and *Universal Prayer;* to the liberal sentiments inculcated in the latter of which he was so much attached, as to have it sometimes sung in his church. How happy should I have been to have had the opportunity of shewing to my countrymen, on my return, the numerous publications, principally historical, for which I was indebted to the liberality of this learned and noble author; but, though unfortunately deprived of this satisfaction, I record, with infinite pleasure, my obligations to him, not only for these, but for various other books which I could not elsewhere have procured. *Two of the works that have come from the pen of the Etatsroed deserve particular mention: the titles, indeed, have altogether escaped my memory, but, if I am not mistaken, one of them was written in the Danish, the other in the Icelandic language, and both treated of the most remarkable occurrences that had taken place in the later history of the country, among which it was peculiarly gratifying to me, as an Englishman, to find, while the author was himself translating some portions to me, how earnestly and how completely *con amore* he bears testimony to the noble and generous conduct of Sir Joseph Banks, impressing, in the strongest terms, upon the minds of his countrymen a sense

* These are in all probability two of the Etatsroed's publications mentioned by Dr. Holland: the one entitled *Iceland in the 18th Century;* the other a translation of the same into Danish, with additions. The former was printed in 1806, the latter in 1808.

of the obligations they owe to him for the unexampled assistance which he afforded to such Icelanders, as had, in the beginning of the present war, been made prisoners in Danish vessels; constantly striving with the utmost zeal to procure their release, and supplying, with unbounded liberality, their pecuniary wants. I must, however, do the Icelanders the justice to say, that there is no need of the assistance of the press to excite a stronger feeling of gratitude on their part, for the benefits that have been conferred upon them by this exalted character; for the eager inquiries that were in every place made after his welfare, by the aged, who still remember his person, and by the young, who know him from the anecdotes told by their fathers and their grandfathers, were a convincing proof of the esteem and veneration they entertain for him: so that, not unfrequently, while wandering over the wastes of Iceland, my heart has glowed, and I have felt a pride, that I should have been ashamed to dissemble, at being able to call such a man my patron and my friend. A short history of the esculent *Fuci*, published by the Etatsroed, has already been noticed at page 46 of this journal. Music, also, claimed a considerable share of the attention, not only of himself, but of all the family at Inderholme, and a large Danish organ occupied a portion of one side of the room. On my expressing a wish to hear some Icelandic music, the whole family came into the library, and, with their voices, accompanied his performance of several sacred airs. I was next entertained with Danish and Icelandic songs, by the Etatsroed's daughter, which she accompanied with tunes upon the *Lang-spel.* This instrument has long been growing into disuse, so that it is now become of extremely rare occurrence, and very few of the natives indeed, excepting the Etatsroed and his family, are capable of performing upon it with any degree of skill. It consists of a narrow deal box, about three feet long, with a wider semi-circular extremity, in which are the sound holes. Three brass wires, or sometimes five, are extended the whole length of this box, and tightened or slackened by means of small wooden pegs, as in our common violin. It is usually played upon with a bow of horse-hair, the instrument itself lying in the mean while upon the table, but the Etatsroed's daughter frequently made use only of her fingers, as with a guitar, in doing which she pressed the end of her thumb alone upon the wires, moving it up and down to produce the different modifications of sound. The annexed representation of the Lang-spel, sketched since my return, from memory, will give a tolerable idea of its form. Von Troil notices another musical instrument, called *Fidla,* which has two strings of horse-hair, and is played in a similar manner with a bow. This, I was never fortunate enough to see; nor did I ever meet with the *Symphon,* mentioned by the same author, and I have every reason to believe that neither the one nor the other has any longer existence; the increasing poverty of the country

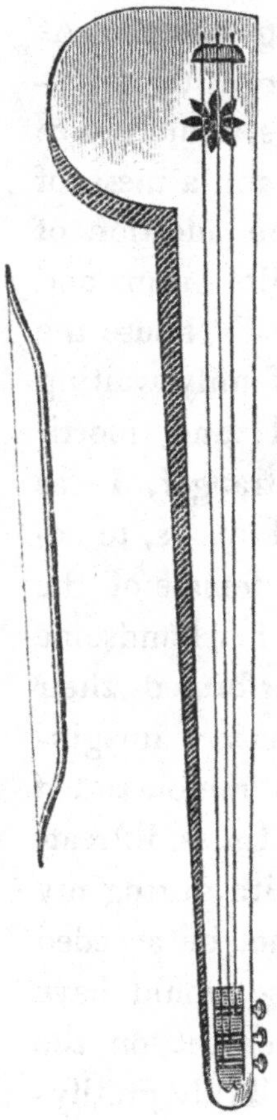

having, probably, been the means of preventing the Icelanders from enjoying the little happiness that they formerly derived from these and various other sources of innocent amusement, which we read of as having been common among them. At about three o'clock we sat down to an excellent dinner of roasted meats, which were eaten with preserved cherries and a mess of the *Rumex Acetosa*, with the addition of waffels, good Norway biscuit, rum and claret. Even in the Etatsroed's house the custom of the ladies of the family waiting at table is religiously observed; and, mortifying as it was to me as a stranger, I was compelled, during the time of meals, to accept of the attendance of the female of the highest rank in the island and her handsome daughter, both of whom performed their parts with the greatest good-nature imaginable. It was in vain that I remonstrated against this relic of barbarous times, intreating it might be dispensed with during my stay: such a request could not be acceded to, for to have done otherwise would have been considered a want of respect on the part of the host to his guest. Truly gratifying was it to observe how much affectionate attention was paid by the younger part of this family to the aged parents of Madame Stephensen, whose father, formerly a syssel-man, although eighty years old, still enjoyed the perfect use of his faculties. Extreme age had deprived the mother of sight, but, though destitute of this comfort, she had the greater one of receiving every possible mark of kindness, that duty or affection could dictate, from her children, who devoted a great portion of their time to bearing her company in her room, and alleviating, by their conversation, her afflictions and infirmities. After dinner I visited the Etatsroed's gardens, which are carefully fenced round by a high turf wall, so as to be, in some measure, protected from the excessive cold of the climate; a precaution that seems to avail but little, for, although in the one adjoining the house, which was laid out in a number of beds, infinite pains had been taken to raise a crop of lettuces, turnips, and potatoes, they all looked in a miserably starved state, and not one came to perfection. Another garden, nearly opposite to the house, was also appropriated to the growth of vegetables, but did not wear a more promising aspect. The ground immediately in front of the Etatsroed's dwelling, though producing a comparatively good herbage, is broken into numerous little hillocks, intersected with rocky divisions, as is almost every where the case in Iceland with the best pasture land, owing, probably, to the treading of cattle between the pieces of rock, which are but thinly covered with earth. Hence to the sea, and for a long way upon the shore, extended a perfectly level tract of country, at one extremity of which, over a little brook, a water-mill had been erected, which was worked by a horizontal wheel, and served to grind corn for the family. This, if I mistake not, is the only one in the island. Some drains, cut by the Etatsroed in an adjoining morass, had greatly improved the soil, and furnished a more copious supply of water to the mill. Were like simple means to be employed in other Icelandic bogs, the greater part of which are admirably calculated for draining, no doubt can be entertained but that the country would be rendered more easily passable, and the increased quantity of fodder, produced in consequence of such an improvement, would be of incalculable benefit to the poor natives. The house of the Etatsroed was but lately erected *, and, as I have before observed, is one of the best, or, perhaps, the very best in the island; yet

* How exactly similar the present mode of building is to that which was in practice upwards of two centuries ago, may be seen by the chapter "De moribus seu communi vivendi ratione," in *Arngrim Jonæ Islandi Tractatus de Islandicæ gentis primordiis, &c.* Indeed, in no part of Europe, I apprehend, have the customs and manners, the language, the dress, and the mode of living, peculiar to a country, been kept so pure, for so great a number of years, as among the Icelanders. "Ut Taciti tempore" (says Arngrim Jonas) "circa annum Christi 120, nec cæmentorum nec tegularum apud Germanos usus, (unde quivis de orbe magis Arctoo judicium faciat,) ita neque postea apud Islandos; sed domus suas ligno et cespite construebant, opere quidem nec momentaneo, nec in speciem deformi; parietes alios solo cespite, alios saxo rudi, cespite pro cæmento adhibito, fiebant; quos postea interius, opere coassato, ut et contignationem ipsam, convestiebant; præcipue in notabilioribus ædificiis. Atque sic tectum cum parietibus ante maturum senium, gramine viridi exterius quotannis enascente, (cespitem namque vivum in tecto et parietibus intellige,) conspiciendum erat. In tecto fenestræ fiebant, raro in pariete; et tecto quidem minus arduo; cujusmodi fuisse fertur apud Orientales tectorum constitutio. Ligna incolis suppetebant, ad littus maris undarum alluvione ejecta; mirando procurationis

its walls and roofs are composed entirely of turf, though so neatly cut, and so well joined, as to present a perfectly smooth and even surface. The doors are ornamented with carved lines, and painted green. The windows, of which there is a double row, are well made, and glazed, and are not in the roof of the building, as in most other Icelandic houses, but in the wall. There are several out-houses for cattle, for provisions, implements of husbandry, drying fish, &c., all which stand apart from the dwelling-house, and are built with equal neatness, and wholly of turf, except the fish-house, which is of wood, formed in such a manner,

divinæ testimonio; cum sylvæ domesticæ, Betulæ tantum, ut existimo, feraces, vastis ædificiis non sufficerent: quæ tamen etiam magno fuere subsidio, cum his quæ incolæ, quoties volebant, ex vicina Norvegia, et fortasse etiam Grönlandia, petebant: utroque enim navigationes annuas longo tempore Islandi habuere. Villarum itaque domus in suo fundo quilibet contiguas fere habebat: præter armentorum stabula, aliquanto intervallo ac ipsis penatibus plerumque sita: item ignaria quædam, non prorsus contigua, ad ignis periculum vitandum: fortasse etiam penuaria quædam quæ solitaria auram et siccantes ventos melius imbiberent." *De regno Daniæ et Norvegiæ Tractatus, p.* 411—413.

that a free passage is left to the air at the same time that the inside is protected from the rain. At no great distance, also, stands the church, a small and neat, though ancient, edifice; and not far from this cluster of buildings rises the steep and rocky front of Akra-fiel, forming a singular contrast with the green plain of Inderholme.

Wednesday, August 2. Immediately after breakfast the Etatsroed, his son a young man of eighteen years of age, and myself, set off for Hvamöre, about twenty miles distance, the residence of the Amptman Stephensen, brother to the Chief Justice, our intention being thence to continue our journey to the hot-springs of Snorralaug, and other remarkable places in the vicinity. This excursion was rendered highly interesting by my having such agreeable companions, and I looked forward with great confidence to deriving from it no small information, as well from the ability of my host to converse with me in English, as from his perfect knowledge of the country, and particularly his intimate acquaintance with its history, in which he is, perhaps, superior to any other person. Our mode of travelling here was new to me, and not a little troublesome. Being about to visit at the houses of persons where the accommodation was good, I had sent Jacob and my guide with the horses and tents back to Reikevig, intending to return thither myself by water; and we all rode the Etatsroed's horses, taking with us a supply of fresh ones to relieve those that carried us the first part of our journey. These, instead of being led, were driven before, without even the precaution of fastening them together; a practice to which some of them that were young and full of spirit did not easily submit; for they frequently strayed away from our proposed course, and gave the Etatsroed's son, Mr. Olav Stephensen *, an infinity of

* In naming his children, the Stiftsamptman, as well as his sons, have abolished the custom, which is otherwise, I believe, very general in Iceland, of calling the child after the christian name of the father, with the addition of *sen* or *son*, to it; thus, the son of the Etatsroed, *Magnus Stephensen*, ought, by this rule, to have been *Magnusen*, to which any christian name might be subjoined. If it had been *Olav Magnusen*, his son would bear the name of *Olavsen*, or rather *Olafsen*, as I believe it is generally written. The females have the addition of *datter* to the christian name of the father.

trouble in pursuing them and compelling them to return into our track; in doing which, he displayed a dexterity and fearlessness in riding that really astonished me, galloping in the most furious manner over the loose fragments of rock. To add to his fatigue, it not uncommonly happened that, when he returned to us after having recovered the horses that had gone astray in one direction, he found those which he had left behind him, and apparently disposed to be quiet, already run off in some different course, so that he had a most tiresome journey. The country over which we passed, after winding round the foot of Akra-fiel * and reaching its opposite side, was altogether

* *Akra,* the name of a parish, means *corn-field,* as the Etatsroed observed to me; and he considered the application of this word to a place in Borgafiord, as a strong argument in favor of the former cultivation of corn in that quarter of the island. From their vicinity to Akra, are also derived the appellation of the mountain Akra-fiel, and of the promontory Akra-ness, and, indeed, we learn from the *Landnama* and *Eigil-Sagas* that Skalagrim, in the beginning of the tenth century, cultivated grain in the southern part of Myrar, and in the neighborhood of the river Hvitaa. On the subject of the cultivation of corn, it is observed in the *Voyage en Islande,* "La

flat and marshy, though many lofty mountains were in sight. In such a place vegetation was of course more abundant than among the rocky and hilly tracts; houses

métairie de Reykholt est le seul endroit dans l'interieur du Breedefiord, dont l'auteur du *Sturlunga-Saga* (1 *B. cap.* 13.) parle aussi avantageusement, en disant que les semailles réussissaient toujours, et que l'on pouvait en tout temps se procurer des farines fraîches dont les habitans se faisaient un régal. Cet historien ne fixe pas précisément le lieu où se faisaient ces ensemençages, mais il parait que c'était près de la métairie où le terrain conservait toujours beaucoup de chaleur par rapport aux feux souterrains, puisqu'il existe ici des sources chaudes et des veines d'eau de même nature, dont les vapeurs communiquent en été aux plantes une humidité fertilisante, et les garantissent en hiver du froid." (*t.* ii. *p.* 83—84.)—If such were really to be the effect arising from the vicinity of hot-springs, either the quantity of corn cultivated must have been very small, or the boiling fountains extremely numerous. But, as the historian has not informed us that this was the reason of the cultivation of corn having been attended with success in his days, it may not unjustly be doubted, whether the circumstance is ascribable to such a cause: for of moisture there certainly seems to be no want in Iceland, and the sudden vicissitudes from heat to cold, which would necessarily arise from a variation in the wind, must, undoubtedly, be prejudicial to plants, which are not capable of bearing the extremes of either. With regard to some of the

also were more frequent than I had elsewhere seen, and the whole district wore an aspect of comfort that seemed to bespeak a greater degree of wealth than is to be met with in any other part of the island. On our left was a bay, or rather arm of the sea, called Borgar-fiord, from which the neighboring country takes its name. At Leera, we stopped at an excellent house, belonging to a Sysselman, who had married a sister of

native vegetables, indeed, a situation like this is not inimical to their existence, but even appears to bring certain species to a greater state of perfection, whilst others are materially injured by it. Of such plants I endeavored to make out a list, but I can at this time, only call to mind, with any degree of certainty, some of the class *Cryptogamia*, which I was particular in examining on the spot, and which I have already taken notice of in two or three places. It may be observed, that a higher degree of temperature in the air extends but a very few yards at any time, in consequence of the steam, and when this latter ascends perpendicularly, the ground receives none of its influence. But how pernicious must be the effect of a westerly breeze, wafting the heated vapor upon the young and tender plant, when followed, as is often the case, almost immediately, by a wind from the east, that drives the steam in another direction, and chills with frost what had been the day before exposed to so much heat!

the Etatsroed, and who would not suffer us to depart without setting before us some coffee, roast mutton, rum, and claret, and forcing us to partake of his hospitality; neither could we prevent him from accompanying us on our way till we had reached a difficult pass upon a mountain, through which he observed that the Etatsroed, however well acquainted with the country in general, would not prove a sufficient guide. As we went along, we observed not far from the road a small turf building, which we found on inquiry to be a printing-office, and at this time the only one in the island. Its distance from Reikevig must necessarily be a source of great inconvenience, and cause considerable delay in the issuing of proclamations and other matters relative to government, to which, indeed, may be attributed in some measure the misunderstanding between Count Tramp and Mr. Phelps; the former of whom made this a plea for the not having published earlier the convention that he had entered into with the captain of the Rover sloop of war. For other purposes it may serve well enough, and its vicinity to the Etatsroed, who furnishes it with more employment than any other person, is of considerable convenience to him, as well as of no small advantage to the proprietor. We now approached the mountain Skardsheidi, which we had to cross in our way to Hvamöre, but we previously touched upon the borders of some brush-wood, which here bears the name of a forest, and is considered the finest in the island. To have entered into the wood would have led us too much out of our intended course, so that I was prevented from judging either of the size of the largest birches, of which it was composed, or of its rank in the scale of Icelandic forests. Of such trees, "if *trees* they may be called, which *trees* are none," as we passed on the outskirts, the tallest did not exceed three feet or four at the utmost, and would scarcely have received a more important appellation than that of bushes in other countries. The sides of Skardsheidi are in many places extremely steep and barren, and its base, from being every where, except in the gullies, wholly environed by an immense wall formed of loose pieces of rock that have fallen from the cliffs above, is thus rendered no less impassable than the parts which are naturally more perpendicular. We ascended through a hollow in one side of the mountain, where the appearance of vegetation, scanty and miserable as it was, induced us to alight from our horses and give up a little time to botanizing. I do not recollect that any particularly rare plants rewarded our researches in this spot, but I well remember how much I was surprised at the extent of the Etatsroed's botanical acquirements, and especially at the readiness and correctness with which he called most of the plants by their Linnæan names. This astonished me the more as his only aid has been a few books, the principal of which is *Lightfoot's Flora Scotica,* and even these he has been condemned to study by himself; there being no individual attached to similar pursuits in the whole island. He consequently expressed great pleasure at being now in company with a person who had made botany one of the chief objects of his attention, and he spared no pains in collecting with his own hands and in directing his son to collect such specimens as I most wished to possess. It was not long, however, before we left behind us all traces of vegetation, and climbed the steeper and perfectly barren sides of the mountain, where we observed nothing remarkable, till we came to the difficult passage through which the Sysselman had volunteered to conduct us. This was a sort of chasm, where a quantity of loose stones and decomposed rock, that had been washed down by the rains, afforded a rugged pathway overhanging a precipice on our right, so narrow as scarcely to leave room for our horses to set one foot before the other. We crossed it, however, in safety, and took leave of our kind friend, who returned to Leera. The higher we ascended the more severe was the cold; and a storm of snow, which we had watched for some time above us spending its rage against the upper part of the mountain, now assailed us, and made us feel still more sensibly the difference between the month of August in Iceland and in England. When we had reached the highest summit, over which we had to pass, a still loftier one, called Honn, of a most extraordinary shape, presented itself to our view. Its figure, from the direction in which we saw it, was almost a perfect pyramid, of a most gigantic size; but what rendered it still more singular was the horizontal stratification, that exactly resembled a flight of steps, each stratum projecting beyond the one above it, and gradually decreasing in width to the pointed extremity. Upon the upper surface of all the lower strata lay a covering of snow, whilst their naked perpendicular sides presented so many black intervening lines: the peak itself was entirely enveloped in snow. So strong an impression has this scene left on my mind, that I venture to lay before my readers the subjoined sketch, made from recollection, trusting it will enable them, better than can be done by a description, to form a correct idea of a place, where the excessive severity of the atmosphere prevented my making a drawing on the spot.

The ground upon which we now rode was so firm and unbroken that, having mounted fresh horses, we galloped for a mile or more on solid rock, till the descent became so steep as to require more caution in our proceedings. On descending somewhat lower, we emerged from the clouds into a clear atmosphere, and had a most extensive prospect of rivers, morasses, mountains, and lofty jökuls; among the latter of which those of Geitland made a most conspicuous figure at no great distance from us, shooting their pointed summits, capped with eternal snow, through the thick clouds that partly enveloped their sides. The mountain, also, called Boula, from its great height and conical figure, formed a prominent feature in the scene: it is likewise deserving of notice on account of the vulgar idea, that there is on its summit (which, by the bye, has proved inaccessible to all who have attempted to reach it) an entrance to a rich and beautiful country; a country constantly green, and abounding in trees, inhabited by a dwarfish race of men, whose sole employment is the care of their fine flocks of sheep *. The Etatsroed particularly directed my attention to four rivers flowing through the centre of as many vallies, each exactly parallel to the other, over the whole

* I think I heard of one or two other Icelandic mountains, concerning which the natives entertain similar notions; but I was not aware that Geitland Jökul was believed to contain such regions of pleasure and happiness, till I observed it remarked in the *Voyage en Islande,* where, at page 168 of vol. 1, it is said, "Les Islandais croyent généralement, d'après d'anciens récits fabuleux, qu'il existe au milieu du Geitland une profonde vallée garnie de superbes prairies, et habitée par une petite peuplade inconnue. Ces habitans vivent de leurs troupeaux, et sont, a ce qu'ils disent, des descendans de brigands et de géans: ils les nomment Ikogarmon dans la *Gamla-Saga,* ce qui signifie homme de bois. Cette fable tire son origine de leur *Grettis-Saga* (chap. 50), où il est dit que Grettis habitait en hiver ce vallon. Qu'à cette même époque, c'est-a-dire vers l'an 1026, il y demeurait un Pâtre nommé Thorir, qui avait deux filles, avec lesquelles Grettis fit connaissance. Que ce vallon est garni de bois et de belles prairies, et qu'il y avait de superbes moutons, bien nourris et de la grosse espece." The ideas concerning fairies and giants, as well as the superstitious notions about the monsters of the rivers and lakes and the appearance of evil spirits, are principally confined to the lower class of people, among whom they are very prevalent. On this subject

of which our superior elevation enabled us to cast a bird's-eye view, though the ridges of mountains that separated them from each other were of considerable height. Their fertility and the abundant supply of salmon *

the authors of the above-mentioned work have made some observations, which, though they may swell the note to an inconvenient length, appear to be well worth transcribing. "On pourrait tres-bien attribuer l'idée qu'ils se font de fantômes et d'esprits malins à la vie triste qu'ils mènent dans ces contrées sombres et désertes, environnées de rochers, de vallons obscurs et de cimetières, puisque c'est là que de tous temps on a eu la folle imagination de croire que les spectres choisissaient leurs demeures. C'est aussi dans la partie septentrionale de l'isle qu'il en est le plus question, tandis que l'on n'en entend presque pas parler vers le sud, où les villages sont plus rassemblées, et où il y a toujours des étrangers, outre les navigateurs qui y viennent passer l'été pour le commerce. Ce qui ajoute encore à leurs affections mélancoliques, ce sont les hivers qui y sont trés-longs, et qui les tiennent conséquemment long-temps dans une solitude attristante; en second lieu, la peur qu'on leur inspire dans le bas âge, et enfin leur état actuel de misère et de pauvreté et leur taciturnité qui n'est éclaircie par aucun amusement."

* Grimsaa, which is one of these rivers, is considered as equal to any stream in Iceland for the quantity of

afforded by the rivers, had been the means of inducing many natives to fix their residence in them. We found the side of the mountain by which we descended more thickly strewed with stones than the summit, and we observed that these stones contained a great quantity of a white or greenish mineral substance, some of which was firmly imbedded in the rock, and some that had fallen from it lay dispersed in many places upon the ground. Of both we gathered many and very fine specimens. On reaching the morass below, we were at no great distance from Hvamöre, the house of the Amptman Stephensen, though, before we could arrive at it, we had to cross several rivers and a very unpleasant country. In our way we passed three or four residences of respectable appearance, the owners of which seemed to possess plenty of good cows and sheep. Hvamöre itself was easily

salmon it produces. "En automne, l'endroit de la rivière, qui est fixé comme guéable, se trouve quelquefois si plein de saumons, que les chevaux ont de la peine à passer, et ne savent où poser les pieds." *Voyage en Islande, tom.* i. *p.* 204.

distinguishable from the other buildings by its superior size and style of architecture, and was to us rendered still more striking and interesting by the numerous and happy groupe of its inhabitants who came out to welcome us to their home. Besides our host, our hostess, and their servants, nine of the handsomest children that I ever saw in the island were present. All these were the Amptman's, and as I happen to have a list of the whole of this family written down in my pocket-book by the Etatsroed, it may be inserted as a specimen of the christian names that are made use of in Iceland:

Stephen Stephensen, Amptman of the Western Quarter of the Island.
Gudrun Stephensen, his wife.
Sigrid Stephensen, his daughter.

Olav Stephensen, Magnus, Peter, Johannes, Stephen,	Sons.
Helene, Ragneidur, Martha,	Daughters.

The customary Icelandic ceremony of saluting each individual, not even excluding the servants, was here a matter of some time, but this being at length gone through, we entered the house, and, after a few cups of coffee, soon found ourselves seated before a dinner of roasted meat, sago-jelly, and waffels. The country round Hvamöre, which is flat and swampy, produces but little that is interesting to the botanist. A *Carex*, however, which grows here in the greatest profusion, deserves particular notice, on account of its utility to the Icelandic farmer. During the course of our ride in the morning, the Etatsroed had pointed out the foliage of the plant in many places, and assured me that it was found the most useful of all the indigenous gramineous tribe; for that it made excellent hay, and the sheep and cows afforded a more copious supply of milk from being fed in pastures where it was abundant. At Hvamöre, acres of ground were uninterruptedly covered with it, and I was here enabled to collect many specimens in flower, and to satisfy myself that it was a species with which I was unacquainted, though approaching very nearly in habit to

C. stricta, from which it differs essentially in being much smaller in all its parts, and in having the spikes remarkably drooping. I had before observed the same plant near Reikevig, and in the neighborhood of Skalholt, but in neither of these places did it flourish so luxuriantly or abound so much as here, where, as just mentioned, the pastures were almost entirely composed of it, and a number of people were now employed in cutting it, and converting it into hay. Another meal nearly similar to the preceding ones concluded the feasting of the day: a thing that would scarcely deserve to be noticed, but for the sake of observing that it was the fourth time in the course of the twelve hours that I sat down to a hot roasted joint of meat: first, when we breakfasted at Inderholme, then at the Sysselman's house at Leera, and now twice at Hvamöre. Each repast, too, was preceded by a glass of rum, and concluded by coffee and chocolate, as well as often by tea.

Thursday, August 3. After breakfast, the Amptman and the Etatsroed, with their two sons and myself, set out for Reykholt, taking with us, as on the day before, horses to relieve those which we first rode. These animals were even more spirited and more disposed to ramble than those we took from Inderholme, and gave for some time sufficient employment to the young Stephensens; but after we had advanced a few miles they became more tractable, and suffered us, when we reached a firm and level country of barren and broken rock, to travel with little interruption at a very quick pace. The first object worthy of notice which we passed was an extensive fresh-water lake, in the centre of which is a small grassy island, and on this, as the Etatsroed informed me, grows a Scotch fir *(Pinus sylvestris)*, diminutive indeed in size, but the only one that was ever seen in Iceland. There was no boat on the water, by means of which I might myself have ascertained this fact, and the distance from the shore was too great for me to be positive how far a small dark spot which I could discern was really the fir in question, or, indeed, a tree of any kind. Some future naturalist may, perhaps, have the opportunity of visiting the little island, and learning the truth of a story, which I believe the Etatsroed only knew from the report of the natives, who are said to have remarked the tree for very many years. Not far from this lake we passed a large heap of stones, much resembling a Scotch cairn, concerning which, Icelandic history is silent, but tradition relates that it covers the remains of some unknown ancient warrior. Our course was nearly north-east, and sometimes close by the banks of the broad river Hvitaa * which, taking its rise from Fiskevatn, empties itself into Borgafiord. In our way we called at the house of a peasant, a skilful workman in wood and silver, of whom I wished to procure some snuff-boxes made of the tooth of the *Walrus,* called by the Icelanders *Rostungr (Trichecus Rosmarus* Linn.), an animal that is not unfrequently cast on shore in the northern part of the island, where the teeth, (on account of their beauty and whiteness, in which circumstances they are quite equal to the best ivory) are eagerly sought after and collected, for the

* This must not be confounded with the stream which bears the same name, and runs near the Geysers from the lake Hvitaa-vatn.

purpose of being converted into snuff-boxes. These are prettily ornamented with silver, variously disposed in fillagree work, and are used by people of rank, particularly by the ladies. Of such snuff-boxes the contents are inhaled in the same way, as of those noticed in the early part of this journal; but, as their shape is different, and I was so fortunate as to preserve (together with my Icelandic dress) one of them which was given me by the Etatsroed's lady, I have thought it deserving of being figured. After crossing the four parallel rivers, of which we had so fine a prospect on our descent from Skardsheidi, we entered Reykholts-dalr, or the vale of smoke; a name the place well deserves from the number of columns of steam that are to be seen rising on both sides of the Reykiadals-aa *. Just at the

* The river of the reeking valley.

mouth of this valley we stopped to rest our horses, near a hill from which five or six fountains were gushing forth, and forming a number of streamlets that poured down along every side of the eminence. These I had already crossed with the help of a stout pair of shoes, and was standing by one of the apertures, when a little English dog, that had accompanied me on this excursion, came running towards me through the scalding fluid, unconscious of the heat of the water. His howling soon made known the pain the poor animal suffered, and so alarmed was he ever after at the sight of water, that it was with the greatest difficulty he could be induced to cross a cold river, nor would he do it till we had gone so far that he was fearful of being lost; so that, for some time subsequent to this accident, we were obliged to carry him over the numerous torrents we had to pass. From these springs, which seem to be what are described in the *Voyage en Islande,* under the name of Tungu-hver *, we passed on to those of Aahver,

* Two of the springs of Tungu-hver have been ascertained by Sir George Mackenzie to throw up their waters alternately in a very remarkable manner. Of

the situation of which is truly remarkable. They issue from a solid rock *, as far as I can

this peculiarity I was ignorant myself, nor perhaps were my friends, who conducted me there, acquainted with the fact. I must refer my readers for a very interesting description and view of this spot to the pages of the gentleman just mentioned.

* The authors of the *Voyage en Islande* seem to consider this rock formed by a deposition from the boiling waters, which, perhaps, may be the case, though the color, which, when I saw it, was almost entirely of a reddish brown, does not exactly accord with their description. There were, indeed, some patches of a whitish substance, that appeared to me to originate in a kind of bolus, thrown out by the water. "Aahver est la seconde source dont on ait connoissance. Sa position la rend remarquable, et l'on peut dire qu'il n'y a pas sa pareille en Islande, attendu qu'elle coule depuis les Thermes de Tungu, an milieu du Reikholtsaa, en prenant vers l'est. La force incrustative de ses eaux a formé peu à peu un rocher qui s'élève à cinq pieds au-dessus de la rivière. Il est d'une telle blancheur, que l'on dirait qu'on l'a endui de chaux; il est constitué d'une concrétion de thermes, qui a acquis la solidité de la pierre. On remarque dans son intérieur, des petits trous, ou, pour mieux dire, des petits conduits courbés d'où jaillissent avec murmure les eaux bouillantes qui partent de son fond. Les bords de ces trous sout colorés en dehors d'un jaune verdâtre, ce qui provient des vapeurs sulphureuses." *t.* i. *p.* 220.

remember, about twenty feet in diameter, standing insulated nearly in the middle of a wide and cold stream, above the level of which it rises to the height of three or four feet. On the summit are two apertures, each of them a foot or a foot and half in width, and from these are almost incessantly spouting little jets of boiling water, which, trickling down on one side of the rock, unite with the cold stream below: there, being carried along by the velocity of the current, they form a line of heated water, the extent of which may readily be distinguished by the little clouds of steam which are continually issuing from it and floating upon its surface. Neglecting other springs of less importance, which, as we journeyed on, were here and there sending up their columns of vapor on each side of us, we hastened forward to the Snorralaug, a place of no little celebrity in Icelandic story, as having been Snorro Sturleson's bath at Reykholt. This is one of the most interesting spots in the country; not merely on account of its numerous hot-springs, and of the superior fertility of its soil over that of most other parts of the island, but also from its having been formerly the residence of the great historian of the north *, from whom the bath derives its appellation. It was here that, in the early part of the thirteenth century, he fixed his abode, after retiring from

* There is a short account of this celebrated man in Mallet's *Introduction à l'Histoire de Dannemarc, &c.*, and, perhaps, I cannot do better than extract a portion of what is there said concerning him, in the words of his translator, from the second volume of the *Northern Antiquities*, pages 22 and 23. "The famous Snorro Sturleson was born in the year 1179, of one of the most illustrious families in his country, where he twice held the dignity of first magistrate, having been the supreme judge of Iceland in the years 1215 and 1222. He was also employed in many important negociations with the Kings of Norway, who incessantly strove to subdue that island, as being the refuge of their malcontent subjects. Snorro, whose genius was not merely confined to letters, met at last with a very violent end. He was assassinated in the night that he entered into his sixty-second year, anno 1241, by a faction of which he was the avowed enemy. We owe all that is rational, certain, and connected in the ancient history of these vast countries, to his writings, and especially to his *Chronology of the Northern Kings*. There runs through this whole work so much clearness and order, such a simplicity of style, such an air of truth, and so much good sense, as ought to rank its author among the best historians of that age of ignorance and bad taste. He was also a

the fatigues of his public duties, and devoted his time to the improvements of his farm and the composition of his numerous works, as well poetical as historical. Here, too, in the turbulency of those barbarous ages, he fell a victim to a midnight assassin, and here he was buried in some part, as it is believed, of the present churchyard, though, the pastor assured me, the precise spot is not known, nor is there a vestige of any monument to lead to its discovery. The only probable conjecture to be formed is, that he lies in that portion of the ground which is still called Sturlunga-Reitur, because, to use the words of Olafsen and Povelsen, "c'est lá que sont enterrés les différens membres de cette famille et quelques-uns de leurs domestiques." The church of Reykholt is of modern date, as is also part of the house of the clergyman which adjoins it; but some ancient rude carvings of figures in wood, which are still very visible upon

poet, and his verses were often the entertainment of the courts to which he was sent. It was, doubtless, a love for this art which suggested to him the design of giving a new *Edda*, more useful to the young poets than that of Sæmund."

the latter over the entrance of the door, and other appearances of antiquity about it, render it probable that a portion of the dwelling has actually existed from the days of the historian. Very near the parsonage is a circular grassy mound of earth, flat on the summit, and evidently, to judge from the sound caused by stamping with the foot, hollow within; but what this formerly was, or to what use it could have been applied, is at present wholly unknown. It has hitherto been suffered to remain entire, from some superstitious notions of the natives, who conceive that it was probably the spot where Sturleson was murdered, and that the disturbing of it would also disturb the manes of their learned countryman. It is far from unlikely that a slight tinge of this superstition affected the mind of the late incumbent of the living, who had just breathed his last before our arrival; since during his life he had constantly resisted the entreaties of the Etatsroed to have the mound opened, a thing that his less scrupulous successor promised should soon be done. At the distance of a few paces from this mound is the Snorralaug, a perfectly circular aperture, about twenty feet in diameter and four or five feet deep, cut in the side of a small hill, and walled round with square pieces of rock, not joined by any cement, but neatly placed together, so as to present a very even surface. The floor is paved with the same materials, and about a foot and a half of the lower part of the wall projects into it, so as to form a bench all round, where twenty or thirty persons may, with more convenience than cleanliness, bathe at once. The boiling fountain in the immediate vicinity, called Skribla *, affords at all times an abundant supply of hot water for the bath, into which it is conveyed through long wooden troughs. By means of a transverse board, moving upon a pivot, the water may be directed to the bath, or turned off to another course, after a sufficient quantity has been admitted; and, for the purpose of reducing the temperature of this water to the wishes of the persons about to bathe, a cold

* Near the source of this spring and attached to the inside of the wooden troughs, I met with many specimens of *Anthoceros punctatus,* flourishing in a very great degree of heat.

stream, from an adjoining spring, is, also, by a similar contrivance, conveyed to the basin, as often as is desirable. By drawing a plug from a small diagonal opening in the bottom of the bath, next the lowest side of the hill, the water, after being used, is suffered to run off, and the place is again fit for the reception of other visitors. In the time of Snorro Sturleson, no doubt, this bath was frequented by the healthy for the sake of cleanliness and luxury, as well as by the sick, for the cure of various complaints; but now it is scarcely ever used, except for the purpose of washing clothes or of bending wood and hoops for casks, and we consequently found it in a most filthy condition. The Sweating-house*, as it is called, situated about a mile from this bath, is another place

* The following mode of heating rooms in use among the Icelanders, as related by Arngrim Jonas, may well be considered as a vapor-bath, and deserves to be noticed here. Speaking of the turf for burning, Arngrim Jonas says, "Quanquam igitur judicarit Plinius miseras gentes, quæ terram suam urerent: nos contra eo nos feliciores ducimus; Deique beneficium hic et alibi agnoscimus, quibus fomes igniarius et cremandi materia non magno constet; qua re ad frigoris intem-

that was erected in former times for persons afflicted with different diseases, but now serves merely for drying the clothes of a neighboring peasant. It is a small turf building erected over a subterraneous boiling stream, which is covered with so thin a stratum of stone that the dry heat arising from it is very considerable, and soon throws into a most profuse perspiration any person who will be at the trouble of creeping into this confined room, as I did, upon their hands and knees, through a narrow and low passage, about five or six yards long. The

periem arcendam, præter alios usus satis notos, incolæ summe indigebant; præsertim hyemalibus temporibus, quibus hypocausta et fornaces in usu, saxo et petris congestæ, per quas flamma facile erumperet; quæ quamprimum ignis vi penitus essent excalfactæ, cumque jam defumasset hypocaustum, frigida camini saxis candentibus aspergebatur; quo pacto calor sese per universam domum efficaciter diffundere solet; qui sic etiam pariete et tecto cæspititio optime conservatur. Memini autem, me balnea publica excalfaciendi similem rationem apud extraneos alicubi observare."—A curious account of this manner of bathing may be seen in *Acerbi's Travels,* where it is said that the natives of Finland have small houses built on purpose for the bath, and that they remain in the vapors for half an hour or an hour in the same chamber, heated to the 70th or 75th degree of Celsius.

closeness of the place, the heat, and the smell of the clothes, soon induced me to retreat, and, having now seen what was most worthy of attention in the valley of smoke*, we turned towards Hvamöre, taking, however, a different route from that by which we had come in the morning. In our way, we stopped a few minutes at the house of a priest of the name of Joneson, where I was agreeably surprised at the sight of a jar of water filled with the charming flowers of *Epilobium frigidum* †, *Fl. Scand.* a beau-

* One would suppose that the quantity of steam must be greater than it really is for it to produce an effect which is mentioned in the *Voyage en Islande.* "La fumée et les vapeurs continuelles qui s'élèvent dans l'air, occasionnent beaucoup de pluies dans le pays : il en tombe même fréquemment dans les plus beaux temps de soleil, mais elles ne durent guères, parcequ'elles ne viennent que d'un nuage qui s'est élevé avec précipitation; il se peut néanmoins que la chûte d'une pareille vapeur de nuages, ne provienne que de la légèreté de l'air." *tom.* i. *p.* 237.

† This plant does not always, as Mr. Salisbury seems to think, grow in maritime situations. The spot where I met with these specimens was at some distance from the sea, and those which I found in the chasm, at the foot of Skoul-a-fiel, could not be less than ten or twelve miles from the coast.

tiful figure of which has been given by Mr. Salisbury in the *Paradisus Londinensis* under the name of *Chamænerium halimifolium.* Our host informed us he had found them on the side of Hvitaa, and I therefore hastened thither, and gathered a number of fine specimens of this splendid plant, the most striking vegetable production of Iceland. I had previously seen it, though in a less forward and luxuriant state. During our stay here, some people who had been requested by the Stiftsamptman to procure me specimens of the minerals of the country, brought me a number of different kinds, among which were several large pieces of *Obsidian* and some fine *Zeolites.* Late in the evening, after a most interesting ride through a comparatively populous and fertile tract of country, we returned to our hospitable abode at Hvamöre, where we rested, and

Friday, August 4.

early the following morning bade farewell to the Amptman's family, or rather to a part of it; for he himself and his eldest son had offered to accompany us to Inderholme, and thence to Reikevig. To vary in some measure our ride, and give us an opportunity of seeing more of the forest at the foot of Skardsheidi, we proposed going round the base of the mountain instead of crossing it. In a short time we reached the shore of Borgafiord, and continued upon a black beach of decomposed rock, as fine as sand, but more firm to the horses' feet, till, finding ourselves in a line with the wood, we turned from the water's side, and, without much difficulty, penetrated to the centre of the forest, where grew the loftiest of the trees that it was composed of, some of which were certainly larger than I had expected to have met with. The tallest, or I am much mistaken, were not less than eleven or twelve feet in height, and measured at the base five or six inches in diameter. In remembrance of the spot, I gathered some of the blossoms of the birch, which were now expanded, and diffused around us an agreeable fragrance that I never thought to have enjoyed in Iceland, while under our feet *Festuca vivipara* and other grasses, with *Silene acaulis* and abundance of the elegant *Polypodium Dryopteris* formed a rich carpet that almost made me forget the desart scenery which was on every side of us. That I might be able to tell my friends on my return to England that I had eaten my dinner in an Icelandic forest, the Amptman spread a cloth, and produced some rum and provisions that he had brought with him for the purpose, of which we partook, protected by the shade of the birch-trees from the rays of the sun, though not from any heat which these rays would have afforded; for the cold was still very severe, and it was but a short time after our sylvan repast, before we had to ride a considerable length of way in the midst of a heavy fall of snow. On coming out from the wood and looking up to a part of Skardsheidi that was below even the height that we had crossed but a few days before, we could clearly discover the currents of water, which we had seen run down the almost perpendicular parts of the mountain, already in a congealed state, and forming so many broad lines of solid ice, the appearance of which, upon the black face of the naked rock, was no less curious than interesting, at such a season of the year. As we approached the shore again, we came among a vast number of huge stones, scattered at various distances about a great plain, so much frequented by eagles, that at one view we remarked no less than five of these birds perched upon the rocks at a small distance from us, and so fearless were they of strangers that I was able to ride within thirty or forty yards of one pair without their offering to move. The unevenness of the country did not admit of a nearer approach, and I had therefore no other means of trying the extent of their self-confidence, except by urging my dog to go up to them, and him they suffered to come within a distance of scarcely more than twenty yards, before his barking at length compelled them to take flight. Both these birds and the ravens do much mischief to the flocks of sheep, particularly in the spring, by carrying away the young lambs. We still continued along the shore, and, in our way, rode at the foot of a most romantic cliff, broken into a variety of picturesque forms, and here and there adorned with tufts of birch and various kinds of willows, while the numerous rills of water, which poured down the sides, afforded nourishment to a thick covering of moss, that added a richness to the coloring. On this grew the beautiful *Epilobium angustifolium*, and I also gathered *Ligusticum scoticum*, though with its flowers scarcely expanded. Soon after, among some loose soil by the side of a river, I found the *Papaver nudicaule* in full flower. Early in the afternoon we reached Leera, where our friend, the Sysselman, who was in expectation of us, afterwards joined our little party to Reikevig. In the evening, as we approached Inderholme, we saw, at a considerable distance, entering the Bay of Faxafiord, a large three-masted vessel, which the Etatsroed supposed might be an American, that was expected to arrive with provisions.

Saturday, August 5. My luggage and horses having, as above-mentioned, been previously sent to Reikevig by land, the Etatsroed, the Amptman, the Sysselman, and myself, accompanied by the eldest sons of the two former, embarked on board a six-oared boat to cross the bay for the same place. In conformity with a custom generally prevalent in Iceland, previously to making an aquatic excursion, all the crew took off their hats and rested a few moments upon their oars, while they offered up a silent ejaculation to Heaven for a prosperous voyage. A light breeze, assisted by the oars, soon carried us away from the shore, and we enjoyed, as we passed along, a fine view of the mountains at the head of Hval-fiord, and even a distant glimpse of Geitland-Jökul. At one time, a large shark rose so near the boat as to cause some little alarm; but the Etatsroed, who was at the helm, quickly made signal to the boatmen to pull more briskly, by which means we soon saw the animal astern of our vessel, where he continued some time in sight, alternately plunging and rising to the surface of the water. A pleasant passage of about twenty miles from Inderholme brought us to the shores of Reikevig, and I here learned that the vessel, which we had observed the evening before entering the Bay, was the Talbot sloop of war, commanded by the Honorable Alexander Jones. She had for some little time been cruising off Iceland, in the course of which she had made a landing on the south coast, and had entered the bay of Havnfiord. From this place the captain had proceeded without loss of time to Reikevig harbor, that he might have an opportunity of ascertaining more correctly the facts connected with a revolution in the government, of which he had heard at the former place, but had received only a short and unsatisfactory account. The consequence of these enquiries was his issuing orders, that the persons, principally concerned in bringing about this change of affairs, should with all possible expedition proceed to England, where a full account of all the transactions was to be laid before the British government. From this time, therefore, my researches in Iceland may be regarded as nearly at an end; and, though various circumstances prevented the sailing of our vessel until the twenty-fifth of August, yet the daily, and sometimes hourly, expectation of being called on board, prevented my making any excursion to a distance from Reikevig. Much of this time was spent in short, but, from the general barrenness of the soil, usually unproductive botanical walks in the vicinity of Reikevig; and a portion, also, in balls and festivities *, as well on board

* These entertainments were common, indeed, on almost every day of the week, but were scarcely ever omitted on a Sunday evening, a custom, I believe, prevalent wheresoever the Lutheran Religion is esta-

the Talbot, as in the town, or in visits to the Stiftsamptman at Vidöe, and to Doctor Clog, the chief physician of the island, who lived at an excellent house at Noes-gaard, where we were sure to meet from him and his lady with a kind and hospitable reception. My memory no farther enables me to continue my journal in any thing like a regular manner, but, even had this been the case, yet still such would be found the uninteresting nature of the events that happened, except, indeed, those political ones that are more fully detailed in the Appendix A, that they could afford but little amusement. I therefore have less reason for regret at having lost this part of my notes, and I proceed to a brief recital of such matter as fell under my own personal observation, but has been omitted to be noticed in the course of my journal; conceiving that it may be of service in adding somewhat to our knowledge of the natural history of the island.

blished. The Icelandic Sabbath commences, according to the Ecclesiastical Laws of the island, at six o'clock on the Saturday evening, and terminates at the same hour on the Sunday.

My inclination rather than my ability leads me in the first place to offer a few remarks on the botany and zoology of the country. In these two great kingdoms of nature, perhaps it would be difficult to find any spot of land, of an equal extent, in a similar degree of latitude, which can lay claim to so small a number of species. The arctic regions of Norway, Lapland, and the Russian Empire, are comparatively rich in these departments; a circumstance most probably to be attributed to their warmer summers, and to the undisturbed state of the soil. In spite of this, however, a botanist, coming from the more temperate climate of Great Britain, will still meet with many vegetable productions that will interest him, such as *Azalea procumbens*, *Cardamine hastulata*, of English botany, *Rubus saxatilis*, *Erigeron alpinum*, *Saxifraga nivalis*, *rivularis*, *cernua*, and *oppositifolia*, *Silene acaulis*, *Veronica alpina* and *fruticulosa*, with many other species, which he has been accustomed to see only on the summits of his loftiest mountains, but which will here be found growing in the plains and vallies, and near the shores of the sea. *Ranunculus lapponicus*, *glacialis*, and *hyperboreus*, *Eriophorum* *capitatum*, *Konigia islandica*, *Gentiana tenella*, *detonsa* (the *ciliata* of *Retzius*), and *aurea*, *Andromeda hypnoides*, *Chamœnerium halamifolium*, *Angelica Archangelica*, *Lychnis alpina*, *Papaver nudicaule*, *Draba contorta* of *Retzius*, *Orchis hyperborea*, *Carex Bellardi*, *Salix Lapponum*, and other plants peculiar to high northern latitudes, together with some, as yet undescribed, will likewise offer themselves for his examination, and afford him a pleasure, of which no one, but a naturalist, can form an idea, as well as what is happily termed by Doctor Smith one of the highest sources of gratification attending upon this and similar pursuits, "the anticipation of the pleasure he may have to bestow on kindred minds with his own, in sharing with them his discoveries and his acquisitions."* But a richer field is open before him in the class *Cryptogamia*. The *Muscologia* of the country is little known, and I am sure, from what I myself found, that many new and rare species would reward a careful search among this tribe, though, like me, he might seek in

* Preface to the *Introduction to Botany*.

vain for the magnificent *Splachna* of the Norwegian and Lapponian Alps, *rubrum* and *luteum*, two plants that I had most earnestly reckoned upon gathering. *Tortula tortuosa*, *Catharinea hercynica* and *glabrata*, with *Polytrichum sexangulare*, the latter always barren, as in Scotland, *Buxbaumia foliosa*, *Dicranum pusillum*, *Hypnum revolvens*, *Silesianum*, and *filamentosum*, *Meesia dealbata*, *Conostomum boreale*, *Splachnum vasculosum* and *urceolatum*, *Trichostomum ellipticum*, *Fontinalis squamosa* and *falcata*, both abundantly provided with capsules, and *Encalypta alpina*, as well as many other mosses, which I cannot with any degree of certainty now call to my remembrance, are met with upon the lava, in the morasses, or in the rapid torrents. Most of the known alpine species of *Jungermannia* are also natives of Iceland, and some new ones, the loss of which I peculiarly regret. Of *Lichens* there are comparatively but few, as, indeed, may reasonably be expected from the extreme scarcity of trees, to which so many of them are exclusively attached; and even the rocky species are far from abounding; the lava, which covers so great a proportion of the island, being eminently unfavorable to the growth of them. On the primitive mountains I observed the more common crustaceous *Lecideæ* and *Parmeliæ,* with some others unknown to me, which the exceeding severity of the weather prevented my examining carefully in their places of growth, and the exceeding hardness of the stone equally prevented my getting specimens of. The perennial snows that cap the higher hills, forbid any of them to grow on very high elevations, as in more temperate climates: in the plains *Bæomyces rangiferinus,* so useful in Lapland as the food of the rein-deer, is found in the greatest profusion and luxuriance; and the singularly elegant *Cetraria nivalis,* which is almost equally abundant, though always barren, makes amends by its beauty for the absence of a greater variety of species. The shores of the island are too much exposed to the most heavy and tempestuous seas, to suffer the more delicate species of submersed *Algæ* to attach themselves to the rocks, and the violence of the surf prevents such as come from more sheltered spots from being thrown uninjured upon the beach. *Ulvæ* I saw none, except *U. lactuca* and *umbilicalis*, and among *Fuci F. ramentaceus* was the only one which came under my observation, that has not a place in the British list. With the larger kinds employed in the making of kelp the rocks every where abound, and I should think that the advantages resulting from the manufacture of this article, which is carried on in Scotland to such a great extent, and has proved so enormous a source of wealth to many of the Hebrides, might, also, with the fostering aid of a benevolent and liberal government be extended to the wretched Icelanders, who have so much greater need of it. A plant, which has been found in Lapland, and which Doctor Wahlenberg, in a letter to Mr. Dawson Turner, calls *Rivularia cylindrica* * of his MSS., is extremely common in the rivers and fresh-water lakes of Iceland, but appears to me to have no nearer an affinity to the genus *Rivularia*, than it has to *Conferva*, to which latter Doctor Roth has lately referred a plant for-

* See page 86 of this work.

merly known under the name of *Ulva lubrica,* with which, in its texture and the disposition of its-seeds, it appears exactly to coincide. It extends from three inches to as many feet in length, unbranched, and, as its name implies, cylindrical, forming an uniform tube, of a pale green color, and thin delicate semi-gelatinous substance, studded all over with darker green seeds, that are almost universally placed in fours, standing in small squares. As I have been fortunate enough to save specimens of this plant, and a drawing that I made upon the spot, I shall, probably, at some future time, take an opportunity of making a figure and more full description of it public. The water of the pools, that have been formed in the morasses, by cutting away the turf for fuel, generally abounds with our common species of *Confervæ,* such as *C. nitida* and *bipunctata;* and a few of our marine ones are found in the basins among the rocks, and upon the sea-shores. But other more interesting species are met with on spots of earth and rock that are heated to a great degree, either by the steam of the boiling springs or by the waters themselves: most of these seem to belong to the Vaucherian genus, *Oscillatoria.* Of *Fungi,* the island can boast but few, except some *Agarici,* scattered in such small quantities, that they are not used for food, and *Lycoperdon Bovista,* which is found every where.

The entomological productions of Iceland are extremely scanty. A very small collection of insects, indeed, rewarded my researches in this department of natural history, and of these there were none that were in the least remarkable for their beauty. Some of the *Lepidopterous* species were new to me, among which I think I had five or six nondescript *Phalenœ.* No *Papilio* or *Sphinx* has ever been met with in the country. Of *Coleopterous* insects, there is scarcely a greater variety; and I saw only a single *Scarabœus,* and a very few *Curculiones* and *Carabi,* most of which, however, to make me amends, were such as I was unacquainted with. I, by mere accident, have still preserved a specimen of an undescribed species of *Coccinella,* which I found killed by the steam of one of the hot-springs at the Geysers: it was the only one of the genus that occurred to me.

The fish of these coasts scarcely at all fell under my observation, so that I have little more to remark upon this subject, than that thirty-three species are enumerated by Mohr, nearly all of which, I believe, are natives of our own seas; but of these almost the only ones that came to our table, were cod, salmon, and the Thingevalle trout. Herrings I never saw, nor are the natives provided with nets for catching them.

Many species of *Molluscæ* frequent the shores, upon which *Medusa cruciata* is often thrown in great quantity, and of a size much exceeding what I ever met with in Britain, not measuring less than a foot in diameter. Shell-fish are far from abundant in the parts I visited, excepting whelks, limpets, and barnacles, which latter, as in England, often incrust large masses of rock, and the *Mytilus modiolus*, which is commonly eaten. Of the more delicate shells I was enabled to gather but a very small number.

The water-birds of Iceland are numerous, most of those which migrate in the winter to our more southern latitudes coming here in the summer to breed, and no doubt many new species may be met with; but other occupations, and the great difficulty of procuring specimens in this country, did not permit me to bestow upon this department the attention I could have wished. I was fortunate enough to procure one or two apparently nondescript species of *Anas;* and a very small kind of *Phalaropus,* with which I was unacquainted, having a body scarcely larger than a lark, was now and then seen near Reikevig: it was probably the *P. glacialis* of Doctor Latham.

I need not here repeat what has already been said in other parts of my journal respecting the few birds I met with in my excursions, nor the particulars I collected about the eider-duck, whose down affords such an important article of commerce; but I have yet mentioned nothing relative to the Icelandic Falcon, which of all the hawk tribe is considered of the greatest value in falconry. This noble bird was, by the older ornithologists, classed among the varieties of the Linnæan *Falco Gyrfalco,* but by Gmelin referred to *F. candidus,* in his edition of the

Systema Naturæ, since which time Doctor Latham and succeeding writers, have raised it to the rank of a distinct species, under the name of *F. islandicus.* It possesses a plumage that varies in the different periods of its existence still more remarkably than that of other hawks; "and hence," as Doctor Shaw observes, "seems to have arisen the wonderful discordance in the descriptions of authors, which have at length amounted to so confused an assemblage of contradictory characters, as almost to set at defiance all attempts to reconcile them." Of the numerous varieties, the white is the most rare, and the most eagerly sought after by the natives; all that are taken of this color being reserved for the King of Denmark, who sets so high a value upon them, and so low an one upon the lives of his oppressed subjects, that a law has been enacted, declaring it death to any man who shall destroy one of these birds. The estimation they are every where held in has induced his Danish Majesty to consider them worthy of being sent as presents to the different crowned heads in Europe, and they have for many years been appropriated to this illustrious purpose. The persons engaged in the catching of them, take them to Bessestedr, where they are examined by the king's falconer, who is sent, annually, for the purpose of procuring a supply of them, and brings with him in the vessel live cattle, to furnish them with fresh provisions during the passage. If the bird, upon inspection, proves not to be of the proper kind or age, it is immediately killed; but, otherwise, there is, according to Horrebow, a reward of fifteen rix-dollars given for a white falcon, and seven for one of the more common varieties. Eagles, as already observed, are abundant in Iceland; and ravens, the favored bird of Odin *, not less so; swans,

* "The *Raven* holds the first rank among the land-birds in the Scandinavian Mythology. We see the use made of them by the chieftain *Floke*. The bards in their songs give them the classical attribute of the power of presage. Thus, they make *Thromundr* and *Thorbiorn*, before a feudal battle, explain the foreboding voice of this bird, and its interest in the field of battle.

THR.

"Hark! the *Raven's* croak I hear,
Lo! the bird of fate is near.
In the dawn with dusky wings
Hoarse the song of death she sings.

shags, corvorants, gulls of different kinds, gannets, stormy petrels, auks, and puffins, are likewise plentiful, and the latter might often afford the natives a salutary and

---

Thus in days of yore she sang,
When the din of battle rang;
When the hour of death drew nigh,
And mighty chiefs were doom'd to die.

THOR.

The *Raven* croaks; the warriors slain
With blood her dusky wings distain;
Tir'd her morning prey she seeks,
And with blood and carnage reeks.

Thus, perch'd upon an aged oak,
The boding bird was heard to croak;
When all the plain with blood was spread,
Thirsting for the mighty dead.

"The *Raven* was also sacred to *Odin*, the Hero and God of the North. On the sacred flag of the Danes was embroidered this bird. *Odin* was said always to have been attended with two, who sate on his shoulders, whence he was called the *God of Ravens:* one was styled *Huginn* or *Thought;* the other *Muninn* or *Memory*. They whispered in his ear all they saw or heard. In the earliest dawn he sent them to fly round the world, and they returned before dinner, fraught with intelligence. *Odin* thus sang their importance:

welcome meal, but that, being destitute of fire-arms, they have no means of killing them. The eggs and the feathers of many of these birds they turn to considerable account. Poultry of all kinds are quite unknown to the Icelanders, except that a few are now and then conveyed to the country by the Danes, who are obliged at the same time to bring with them a sufficient supply of necessary food for their support, the island itself furnishing none.

Indigenous quadrupeds, likewise, as has already been remarked in a previous part of my journal, are wholly wanting.

Among the domestic animals in the island, the dog deserves the first place, not only as the companion and solace of the natives as well as the guard of their houses, but as being of

---

"*Huginn* and *Muninn*, my delight!
Speed through the world their daily flight;
From their fond lord they both are flown,
Perhaps eternally are gone.
Though *Huginn's* loss I should deplore,
Yet *Muninn's* would afflict me more."

Pennant's Arctic Zoology, Introduction, p. 72.

essential service in their agricultural pursuits, by keeping the horses from eating the grass intended for hay, and by collecting the sheep scattered over the mountains, and driving them to the milking-places. Hence they abound throughout the country, and few huts are unprovided with one or two of them. The *Fiaarhuundar* of the Icelanders *(Canis islandicus* of some authors), if it has not sufficient characters to rank it as a species, is at least a very strongly marked variety; differing in many points from any of the dogs I have elsewhere seen, but most nearly approaching the figures and descriptions that are given us of the Greenland dog. It is rather below the middle size, well proportioned in its parts, having a short and a sharp nose, much resembling that of a fox, and small erect pointed ears, of which the tips only, especially in the young animal, hang down; the hair is coarse, straight, and thick, very variable in color, but most frequently of a greyish brown; the tail long and bushy, and always carried curled over the back. The following circumstance concerning the dogs in Iceland is so extraordinary, that, had I been the only person who witnessed it, I should scarcely have ventured to relate the anecdote; but my scruples are removed, as, so far from this having been the case, I was not even the first who saw it; for Mr. Browning, an officer of the Talbot, whose ill health confined him to a room on shore, called my attention to it, by more than once remarking to me that he had, from his window, in the morning of several successive days, observed at a certain hour a number of dogs assemble near his house, as if by a previously concerted arrangement, and, after performing a sort of sham fight for some time, disperse and return to their homes. A desire to be an eye-witness of so singular a fact, led me to go to this gentleman's room one morning, just as these animals were about to collect. The spot they frequented was across the river, which there are but two ways of passing from the town without swimming; the one a bridge, the other some stepping-stones, each situated at a small distance from the other. By both these approaches to the field, the dogs belonging to Reikevig were running with the greatest speed, while their companions of the neighboring country were hastening to the place of rendezvous from other quarters. We counted twenty-five of them, not all of the true Icelandic stock (the *Fiaarhuundar)*, but some of different kinds, which had probably been brought to the country by the Danes; and I presume it was one of these, much larger and stronger than the rest, who placed himself upon an eminence in the centre of the crowd. In a few seconds, three or four of them left the main body, and ran to the distance of thirty or forty yards, where they skirmished in a sort of sham battle; after which, one or two of these returned, and one, two, or three others immediately took their places: party succeeding party, till most, if not all, had had their share in the sport. The captain remained stationary. The engagement was in this manner kept up by different detachments, the dogs continuing their amusement in perfect playfulness and good humor, though not without much barking and noise, for about a quarter of an hour, when the whole of them dispersed, and took the way to their respective homes in a less hasty manner than they had arrived.

Four species of *Phoca*, are noticed by Mohr, in his *Natural History of Iceland*, as being found upon the shores of that island. The common seal, *Phoca vitulina*, is extremely abundant, and is killed by the natives for the sake both of the skin and the oil: of the former they make their shoes and thongs, as well as bags for various purposes, and an excellent kind of portmanteau, which is composed of nearly the whole hide, with very little alteration, except the cutting away of the head and legs; each extremity being closed by a flat and circular piece of wood, while the opening made for the purpose of skinning the animal is left for the admission of different articles that may be wanted during a journey. It is then fastened behind the saddle upon the horse, as a cloak-bag.

The horses of the Icelanders are small, seldom rising above thirteen hands high, but strong, and though, for want of a proper supply of food, generally in a miserable condition during the winters, when they for the most part are kept among the mountains to procure their subsistence as they can; yet, in the summer, when grass is plentiful, they are well furnished with flesh, and, if not worked too hard, will even grow fat. Every Icelander keeps his riding-horse, and many of the peasants have, also, from fifty to sixty, or even a hundred, others for burthen. These of course are useless in the winter, but, as soon as the fisheries commence, or the season for trade summons their masters to Reikevig and other ports, they are all called into employ, and, if the journey be long, the natives with their tents and families lead, like the Nomades of old, a truly wandering life for nearly the whole summer, subject to no restraint, but taking up their abode wheresoever a pleasing spot or a supply of grass for their cattle invites them, and neither shortening nor protracting their periods of rest, by any other consideration, but their own inclinations: truly happy, if the happiness of man consist in his will being his law! No wheel carriages of any kind can be made use of in the island: every thing is therefore transported upon horses, which renders a number of these animals of the greatest importance to those Icelanders who live at a distance from the coast. It is stated by Povelsen and Olafsen that the price of a horse in their time (about 1750 or 1760), varied according to its goodness from six to eight rix-dollars, and that it was rarely known that one sold for so much as ten or twelve. Now, however, they are so considerably enhanced in price, that I could not buy a good riding-horse for less than thirty rix-dollars, and I have even known persons refuse one hundred for a very handsome one. Sir George Mackenzie* was in this respect more fortunate; for he states that the baggage-horses he bought for his tour to Snæfell Jökul, a journey of three or four hundred miles, cost from eight to ten rix-dollars each, and those for the use of himself and friends about twelve. He adds, however, immediately after, that these were by no means of the best description of riding-horses, but that an exceedingly good horse might be procured for twenty or thirty rix-dollars, a sum according to the rate of exchange at that time equivalent to two or three guineas.

The cows are likewise small, and are seen both with and without horns, but generally

* *Travels in Iceland, p.* 133.

in the latter state. Almost every peasant has five or six of them, though he can seldom preserve the whole through the winter, on account of the miserably scanty supply of hay, which it is alone in the power of the Icelanders to collect from their pastures, to maintain their stock during the long continuance of the season when the ground is covered with snow. It has been well observed, on the subject of this inestimable animal by the writer just quoted, that it affords the principal source of wealth, comfort, and subsistence to the natives. "Milk is almost their only summer beverage. Whey becomes a wholesome, and to them a pleasant, drink in winter. Even fish itself, their primary article of food, is scarcely palatable to an Icelander without butter; and curds, eaten fresh in summer, and kept through the winter, yield the most precious change of diet, both for health and pleasure, which he enjoys. A cow on the farm of the Amptman Stephensen, we were assured, gave regularly every day twenty-one quarts of milk. Their value is well known and appreciated by the Icelanders, who take the greatest care of them throughout the winter, and seem to shake off their habitual listlessness, while employed in gathering in the hay that is to support them through the inclemencies of that season." In years of extreme scarcity * the poor beasts are fed with dried fish cut small; and the authors of the *Voyage en Islande* state it as a fact, that the inhabitants of the islands of Breydefiord have even been reduced to the necessity of nourishing them with dry turf. A cow sells, according to the quantity of milk she gives, at from ten to twenty, and thirty rix-dollars.

I have already made mention in one or two places of the Icelandic sheep, and have particularly noticed the smallness of their size, and the general coarseness of their wool. This latter is never shorn, but is either plucked by hand, or suffered to fall off in the early part of the summer. The first

* The last winter (of 1810) has been peculiarly severe in Iceland, and the cattle reduced to the greatest distress for want of food in almost every part of the country. I have been lately informed by Capt. Liston, who has returned from Reikevig this summer, that during the previous winter, even in the town itself, all the horses and cattle were fed with chopped fish.

wool is extremely fine and short, but, as the winter approaches, a longer and coarser kind is mixed with it, which is said, by writers on Iceland, to be employed in making buttons and garters at Copenhagen, and to be sold for a manufactory of camel's hair. The finest of the Icelandic wool is selected by the merchants at Copenhagen, and considered far superior to the best that Zealand produces. In the neighborhood of Reikevig, sheep sell at from three to four dollars a head, but in the interior of the country they may be bought at very much less. I have paid one dollar for a good sheep, and the peasant has been more than satisfied. For a lamb of a moderate size, two marks (1*s.* 4*d.*) is a fair price. These animals seem to be fond of various species of sea-weed, which they eagerly devour at the ebb tide upon the shores; but it is only when they are greatly distressed for other food, that the natives give them the refuse of the stock and wolf-fish. They are also said at those times to feed them with small narrow pieces cut from the belly of the shark.

Goats, which were formerly abundant in the island, are now but seldom seen, and, I believe, are principally confined to the northern and eastern parts of the island, where some farmers keep small flocks of them. To judge from the skins that I procured of two of these animals, they arrive at a large size, and, from their extreme hardiness, I should have supposed they would have answered well to an Icelandic peasant. Rein-deer I have already noticed as having greatly increased in the mountainous and less frequented districts; and there is reason to hope that at some future period they may be of real importance to the Icelanders. Hogs are no where to be met with, the country unfortunately furnishing no food for their support.

The dark nights which immediately preceded our departure from Iceland gave me an opportunity of seeing the Aurora Borealis in a degree of perfection unknown to the inhabitants of milder climates, though, according to the report of the natives, it was even then very much inferior to what it appears in the still darker and longer evenings of winter. I do not at all recollect observing the light occupying any of the northern hemisphere, but various parts of the east, west, and south were frequently illuminated.

Its color was of a paler yellow than what I had been accustomed to see either in England or the north of Scotland, and its figure most variable; sometimes extending in one narrow line apparently half-way across the heavens; then rapidly expanding in width and contracting in length, altering in form and brilliancy every moment. Sometimes, too, these meteors are confined to one single spot, while at other times they are seen in many different parts at once, but shifting their situations every instant. Upon this subject, Povelsen and Olafsen, whose opportunities of making remarks were so greatly superior to mine, at the same time that they confirm my observation how extremely variable the Aurora Borealis is in Iceland in its form and situation, add, that it is not less so in the periods of its appearing. They say it is rare to see it illuminating the horizon without at the same time being sensible of an evident unsteadiness in it; and that it often exhibits the various hues of red, yellow, green, and purple, now flickering with an undulatory motion, and now shooting out into lengthened straight lines. (en forme de fusées.)

I forbear to speak of the mineralogy of the island, because my ignorance of that important branch of natural history would prevent my being able to offer any remarks farther than I could collect from other authors. Few countries, perhaps, present so interesting a field for the geologist.

While waiting for the sailing of the ship, one of my little excursions in the neighborhood of Reikevig led me to Bessestedr, about eight or nine miles distant, a place that was for a long time the residence of the governors of the country, but is now only remarkable for having one of the neatest churches I any where saw, and a Latin school, the only one in the island. On this account, I may be the more readily allowed, in addition to what I have to offer from my own observation, to enlarge upon its history, with which I am furnished by Mr. Jorgensen, who accompanied me in this expedition. The building itself is of stone, and tolerably good, having of late undergone considerable reparations, but the filth within can scarcely be exceeded by the worst of the poor-houses in our country. A staircase, encrusted with a thick coat of dirt, led us into a bed-room, which was even in a still worse state of nastiness. The bedsteads were merely elevated frames, or rather boxes, of wood, filled with heaps of the sea-grasswrack, *Zostera marina,* gathered from the neighboring shores, which, with the exception of two extremely coarse woollen rugs, constituted the sole furniture of the beds. In each of these, three boys slept, so that there were only eight beds for twenty-four scholars. The excessive closeness of this place, and the horribly offensive smell, which was scarcely bearable, urged us to hasten forward to the library, a small and dirty room, in which a number of books, principally in Latin and Greek, many of them on theological subjects, were lying in great confusion. The college nevertheless, possesses able teachers in the two Jonesons. There were, originally, two schools of this description in Iceland, the one at Holum, the other at Skalholt; in the former of which sixteen scholars, and in the latter twenty-four, were educated, and provided with board, lodging, clothes, and every necessary, and the expences were defrayed out of the revenues of the estates belonging to the two episcopal sees. In the year 1785, the king ordered the estate belonging to Skalholt to be sold by auction, and the money to be deposited in a chest, called Jordebog's Casse, from which the bishop and teachers were thenceforth to receive their annual salaries. The school was then removed to Bessestedr, and each of the scholars allowed a yearly stipend of twenty-five rix-dollars, in lieu of clothes, food, washing, &c. In 1801, in a similar manner, the estate belonging to Holum was sold, the money paid into the same funds, and the two schools incorporated into one, at which, however, even in the first instance, no more than thirty boys were educated; and that number was soon after reduced to twenty-four as it now remains. This reduction was, in all probability, caused by the increasing prices of provisions, which rendered it necessary that an additional stipend should be paid for each boy; and the allowance was accordingly raised to forty, and afterwards to sixty, rix-dollars; but even this is far from being found sufficient. Their food is almost as ordinary as that of the poorer peasantry, consisting principally of dried fish, sour butter, and now and then mutton. Among the improvements, which it was Mr. Jorgensen's intention to have made in the island, had he been permitted to have retained his office as governor, that of bettering the miserable condition of the scholars at Bessestedr was not the least meritorious, or of the least importance. He had appointed Bishop Videlinus, Provst Magnussen, Assessor Einersen, and himself, directors of the school, and ordered that one or other should attend for a few hours every day. A thousand dollars were immediately advanced for the purchase of more wholesome provisions than the boys had been accustomed to, and the house was directed to be thoroughly cleaned and white-washed: during the time required for which, the scholars were all sent home to their parents, who were allowed sufficient money for their daily maintenance. A quantity of linen, cloth, and other necessaries were purchased to provide the boys with clothes, shirts, &c., and proper bed-places were ordered to be made, so that only one lad should sleep in each. Bed-clothes and sheeting were prepared, and every thing done that could tend to the health, the comfort, the happiness, or the cleanliness of the boys: additional salaries, also, were given to the lecturers and teachers. Whether or not these alterations were continued after Mr. Jorgensen left the island, I cannot pretend to say, but, in all probability, the school at Bessestedr, like other things, went on in its old course. We are not, however, to judge of the state of literature and learning in the island, from the small number of boys who receive a classical education at the school of Bessestedr. Many obtain a very considerable share of knowledge in the Latin and Greek languages, and become good scholars, who have never entered its walls. An attachment to reading and study, if not a necessary consequence of the long winters, which for many months immure the natives almost entirely in their houses, is certainly materially increased by that circumstance; it being impossible to find the comforts of society in so scanty a population, and the enjoyment derived from literary pursuits being the only resource left them against the tediousness of so dreary a confinement. The sagas, or traditional histories of the country, are well known to the lower ranks of people, and the comparatively few, who are not able to read, commit them to memory; the delight of a winter's evening in Iceland being for the old to repeat them to their infant posterity, by which means they are continually handed down from generation to generation, as the Poems of Ossian among the natives of the Hebrides. That learning in Iceland has been in a state of decline for some centuries past is allowed even by the present inhabitants; but there are still among them able scholars and great theologians who would do honor to any age or country. Poetry is to this day much cultivated, and it is customary, as often as strangers of rank visit the island and confer upon it, or upon its inhabitants, any signal benefit, to celebrate their actions in poems written upon the occasion. The liberality of Sir Joseph Banks, which I have so repeatedly had occasion to mention, has enabled me to offer to my readers * some of their Latin versions of poems of this description, together with one or two spe-

* See Appendix D.

cimens of their epistolary composition. How little this poetical talent has suffered by a lapse of nearly forty years, since the period of Sir Joseph Banks' visit, will be seen by the last article of the same Appendix, where Captain Jones has kindly permitted me to insert the ode written and presented to him, by an eminent scholar of the present day, Finnur Magnusen, which has been already noticed at page 41 of this journal.

Previously to our departure from Iceland, another change in the government took place, which will be more fully detailed in the Appendix A., before alluded to; yet, nevertheless, as I have, in the early part of my narrative, noticed the seizure and deposition of Count Tramp, and the elevation of Mr. Jorgensen to the dignity of Stiftsamptman, it may not be improper here to add, that an agreement was now entered into between Captain Jones, Mr. Phelps, and the principal Icelanders, by which it was settled that the former government sould be restored, and that it should be held responsible for the persons and property of all British subjects. It was still farther stipulated, that the island should not be suffered to be put into a state of defence; that the convention with Captain Nott should be in full force throughout the country; and that, till definitive orders were received from the British government, the chief command should be vested in the hands of the two persons who were next in authority to Count Tramp, the Etatsroed Stephensen, and his brother, the Amptman of the Western Quarter of Iceland. These affairs having been brought to a conclusion by Friday the 25th of August, the Margaret and Anne and the Orion were finally ordered to prepare to weigh anchor in the afternoon of the same day. In the former we had, in addition to the party we brought out with us, Count Tramp, who was to go to England as a prisoner of war, his secretary, and Lieutenant Stewart of the Talbot, charged with dispatches from Captain Jones to the Admiralty. The Danish prisoners belonging to our prize were divided in the two vessels, and Mr. Jorgensen, together with a few English, sufficient to protect the ship, embarked on board the Orion. At about four o'clock in the afternoon we were both under sail, but

Friday, August 25.

with so little wind that it was evening before we were quite clear of the small islands of Akaroe and Ingle, and the same weather continued till noon of the next day, when a breeze
Saturday, August 26. springing up we soon bade farewell to the Orion, which we now left far behind, observing to each as she faded from our sight that we should never see her again; and, finding we were not near enough to the land to go through the most usual and the safest, as well as the shortest, passage between Cape Reikanes and the first of the rocks called the Fugle Skiers, we made our course between the second and third of them. I believe not one of our little party left Iceland with feelings of regret. The weather, which had at the best been unfavorable, was now daily growing worse, and not only rendered our longer abode in the island disagreeable, but threatened us with a dangerous passage homeward: the nights were rapidly lengthening, and time hung heavily upon our hands: it was impossible to forbear contrasting the wretchedness and poverty of every thing about us with the comfort of our happy homes; and, in addition to these and similar considerations, our stay at Reikevig,

had been in many instances rendered unpleasant by political squabbles, by commercial misfortunes, and, above all, by the ill conduct of some of the persons employed by Mr. Phelps in an inferior capacity. A delightful wind now added to our happiness, and we congratulated each other on the prospect of a short and prosperous voyage to our native shores: but the next morning

Sunday, August 27.

far other ideas crowded upon our minds, when about six or seven o'clock we were awakened by a smoke and a strong smell of burning, that issued from the different hatchways, especially from that in the fore part of the ship, and left us no room to doubt but that the vessel was on fire, and that the flames would soon burst out! No one who has not been in a similar situation can have an idea of what we felt. We were than twenty leagues distant from the nearest shore, a barren and inhospitable coast, and the wind was blowing from that quarter, so that to gain even this was impossible. We were also unprovided with boats sufficient to have contained one half of our crew, nor could any boats have assisted us in such a tempestuous ocean; so that our joy was inconceivable and our a stonishment scarcely less so, when, but a few minutes after the discovery of our misfortune, a distant sail was detected, which, improbable as it seemed to us, we knew could be no other than the Orion. It proved that, contrary to the orders expressly given for her to follow our track till we had cleared the rocks, Mr. Jorgensen had insisted upon the master's taking that short course which we had considered too perilous, and steering between the Cape and the first of the Fugle Skiers; such being the only chance of his not being compelled entirely to quit our company. This he had effected in safety by his courage and superiority in seamanship, and having by this manœuvre gained a sufficient length of way to compensate for the inferiority of his sailing, he was enabled to save the lives of the whole ship's crew, who must otherwise inevitably have perished. After having put about our vessel, and come sufficiently near, we hoisted signals of distress, upon which the Orion crowded all her sail, and in about two or three hours Mr. Jorgensen himself came on board. The fire had by this time so much increased, that it was found necessary to have all the boats in readiness to convey the people to the Orion. Every precaution was in the mean while used to suffocate the flame with wet-swabs, sail-cloths, &c., and thus at least to retard the disaster; but all to no purpose. We so plainly saw our situation, that it was but a little time before the whole of us had left the Margaret and Anne, except a few who remained to cut open the decks and make a last effort by throwing down water to extinguish the flames: such, however, was the ascendency they already had gained, and such the volumes of smoke and fire which instantaneously burst forth, that delay only endangered the lives of the men, and it was found necessary almost immediately to abandon the attempt and give up the vessel to her fate. By twelve or one o'clock every living thing, not even excepting the sheep, cats, and dogs, was secured, but of our property it was impossible to save any thing, excepting only a very few articles that were with us in the cabin; for the fire, at the time of its first discovery, had taken hold of the place in which every thing most valuable was kept. We were but too happy to escape with our lives, and with the clothes upon our backs, and even for this we are in no small degree indebted to the extraordinary exertions of Mr. Jorgensen, at a time when nearly the whole of the ship's crew seemed paralysed with fear. He, too, as would be expected by all who knew his character, was the last to quit the vessel. Just at this time the wind, which had blown fresh, suddenly fell, and we were compelled by the succeeding calm to be the near and melancholy spectators of the destruction of a ship of five hundred tons burthen, with all her sails set, and a cargo principally consisting of oil and tallow, the whole worth not less than £25,000. The flames first seized the sails and rigging of the foremast, which they soon destroyed, and communicated to those of the main and mizen masts, enveloping the whole in one general conflagration. Shortly afterwards they subsided, leaving the naked masts here and there on fire; but when the tallow and oil boiled over and ran in wide cataracts of fire down the sides of the vessel, blazing over every part of the hull, the scene was awful beyond description. The clouds of smoke, greater by far than those of steam from the largest eruption of the Geyser, rose to an almost inconceivable height in one steady column, which was only at intervals disturbed by the discharge of one or other of the guns, or by the falling of the masts. It was not long before the timbers of the vessel were destroyed, but the copper bottom continued floating about, like a great caldron filled with every thing that was combustible in a liquid and blazing state, till the sad spectacle was concealed from our view by a dense fog at four or five o'clock in the afternoon, when with a fairer breeze we steered back for Reikevig, the Orion not affording accommodation for so many people as were now on board, nor being furnished with provisions enough for a voyage to England. It had been whispered among our crew, previously to their leaving the Margaret and Anne, that some of the Danes had probably set fire to the vessel, and this suspicion was now confirmed even by their own confessions. Two of them, therefore, who were most strongly suspected, were put in irons, and the beds, &c., of those belonging to the Orion searched for any combustible matter by which a similar act of villainy might here be committed. The result of this search was, that a large piece of touchwood was found concealed under one of their hammocks, and it was ascertained that it was with some of the same substance that one or two of the Danes, in the Margaret and Anne, went down the fore hatchway at about ten o'clock on the Saturday night, and set fire to the wool, which, owing to its slow mode of burning, was not discovered till the following morning. In the Orion, which was now on many accounts so uncomfortable, we passed but two nights; for on the Tuesday morning we came to anchor in Reikevig Bay, where we landed the whole of our prisoners, except the two in irons, who were received into the Talbot, and in two or three days the Orion again set sail for England with Mr. Phelps and Mr. Jorgensen. Count Tramp and myself were left behind: the former at his own request was received on board the Talbot, and I was likewise invited in the most handsome manner by Captain Jones to take my passage to England in the same vessel, he knowing the poor accommodations that the Orion afforded, and justly supposing that I should be more comfortable with him. I gladly avail myself of this opportunity to acknowledge with gratitude the many marks

Tuesday, August 29.

of attention, and the uniform kindness which I received, both from him and the whole of his officers, not only during the voyage, but also previous to our final departure, which was delayed for a week after our return. They were unceasing in their endeavors to afford me every accommodation and assistance in their power, of which I stood greatly in need, and to make me forget what I had suffered: nor must I pass in silence the kindness of the principal Icelanders, who pressed upon me with congratulations for my safety; especially the Etatsroed and the Bishop, both of whom offered to do whatever they were able, to repair the losses I had sustained, and have since given unquestionable marks of the sincerity of their offers, by having recently sent me collections of plants * and minerals. The Bishop, in a letter now before me, says, "Cum gravissimo sanè dolore calamitatem vestram accepi! Paulsonius noster tibi plantas quas orientalis insulæ plaga hoc tempore producit

* This collection contained one or two plants not before known as natives of Iceland, which I have therefore inserted in the list of the vegetable productions of the island, contained in the Appendix E.

exhibebit. Si quid in meâ potestate erit, quæ amisisti aliquo modo restituere, fac, jube, hoc grato fungar officio." I did not, however, then avail myself of his civility, but spent nearly the whole of my time on board, for there was, indeed, little that could afford me amusement on shore; as it was too late in the season to replace my lost collection of the vegetable productions of the island, neither had I materials to enable me to preserve any subjects of natural history: books, too, were not to be procured without much time and trouble; drawings required still more; and my inclination, it may be well imagined, was not favorable to any of these attempts.

On the 4th of September we once more left these unfortunate shores. It was the captain's intention to have entered a port on the eastern coast of Iceland; but, after beating about for several days within sight of the snow-mountains near the south coast, making at the same time but little progress, we directed our course straight for England, proposing in our way to touch at the Ferroe Islands. With an excellent breeze and fine weather we entered the cluster, but had barely time sufficiently to admire the immensely steep rocky precipices, and strange shapes of the Great and Little Diamond and others of these singular islands, before the clouds rolled down their black sides, and in a very short space of time enveloped us in so thick a fog, that it was considered imprudent to endeavor to attempt to enter the port of Thoreshavn. We accordingly made all sail to clear the islands, which was not fully accomplished when we had the misfortune to lose our foremast, and in consequence of this loss to pass a night of painful anxiety in a severe storm, our vessel almost unmanageable and in continual danger of striking upon some of the neighboring rocks. The excessive darkness of the night, the dreadful heaviness of the sea, and the pelting of the rain, added to the unpleasantness of our situation; and this was still farther increased by the necessity we were under of cutting away our first jurymast, which was in fact no more than the stump of the old one newly rigged, and proved to be unsound. Another was with difficulty set up, and by dawn the next morning we happily found ourselves clear of the islands. The storm, however, continued with almost unabated violence for two nights, in one of which our jolly boat was washed away from her lashings, and broken in halves by the violence of the sea. On the 20th of the same month we thanked God on finding ourselves safe at anchor in Leith Roads.

END OF THE FIRST VOLUME.

Printed in the United States
By Bookmasters